HOW MODERN MEDICINES ARE DEVELOPED

Edited By
FRANK H. CLARKE, Ph.D.

FUTURA
PUBLISHING COMPANY, INC.
1977

Published by
Futura Publishing Company, Inc.
295 Main Street
Mount Kisco, New York 10549

L.C. #: 76-41063

ISBN #0-87993-073-X

FOREWORD

The discovery and development of new medicines are complicated processes. Modern pharmaceutical research is composed of an infinitesimal amount of romantic inspiration and an overwhelming amount of perspiration. It is, above all, a team process. At its best, drug development is an almost continuous interplay between chemists, biologists, statisticians, clinicians, toxicologists, engineers—and businessmen. The process is difficult, time-consuming, and increasingly expensive.

This book conveys well to the reader how this process works. The process is a remarkable scientific and social effort, and all too poorly appreciated by the lay public, including the politicians and media folk, and by the scientific community, including most of academia. It is a story in which the drug industry can take justifiable pride. Anyone reading these chapters will appreciate the diversity of talents needed to make the whole thing go. The reader will also realize how mischievous (and indeed surrealistic) are those suggestions which surface from time to time to the effect that all drug development should be left either to the universities or the government. Nothing could be more disastrous, although *collaboration* between academia, government, and private industry is certainly in the public interest.

The stories in this book deserve a much wider audience than was represented at the symposium which originally evoked them. Drug development is too important to all of us for its details to be known only to a relatively few people in industry, no matter how competent and well-intentioned they are. Many of the decisions that affect drug development these days—

almost always with adverse affect—are in the hands of people who do not understand the process. Let us hope that they read and appreciate the message of this book.

Louis Lasagna, M.D.
Department of Pharmacology & Toxicology
University of Rochester
School of Medicine and Dentristry
Rochester, New York

ACKNOWLEDGMENT

It is a pleasure to express my appreciation and thanks to William C. Cray, vice president and Pamela A. Taylor, publications editor of the Public Relations Division of the Pharmaceutical Manufacturer's Association for their excellent editorial revision of the manuscripts and for the index; to William Cray for making arrangements with the publisher, and to John Adams, vice president, and C. Joseph Stetler, president of the PMA, for helping to make this project possible. This project was sponsored by the Committee on Research Information and Professional Relations of the Research and Development Section of the PMA.

The generous support of the Pharmaceuticals Division of CIBA-GEIGY Corporation is gratefully acknowledged. Special thanks are due to Karl J. Brunings, vice president, for many helpful discussions; and to Richard Brown of the Art Department and Joseph H. Romanchik of the Photography Department for assistance with several of the illustrations and to my secretaries, Anne Porto and Dorothy Vivian, for typing portions of the manuscript.

Sincere appreciation is expressed to the authors and publishers who graciously gave permission to use previously published illustrative material. Individual credits are acknowledged in the figure legends.

The helpful assistance of Steven Korn, Jacques Strauss and Elaine Lipton of the Futura Publishing Company is also gratefully acknowledged.

Frank H. Clarke, Ph.D.
Pharmaceuticals Division
CIBA-GEIGY Corporation

CONTENTS

CHAPTER 1

Introduction

Frank H. Clarke

C.J. Cavallito, writing in a recent issue of *Drug Development Communications* has defined development as "the conversion of discovery to something useful in the broad sense of utility." In *How Modern Medicines Are Discovered* the process of drug discovery was illustrated in a series of chapters written from the vantage point of chemists actually involved in the innovative research process. They illustrated the teamwork that is involved between research chemists and biologists to provide a "lead" compound with potentially useful therapeutic properties. The following chapters illustrate some of the factors that bridge the gap between discovery and delivery of useful therapy. This research can become an order of magnitude more expensive than that involved in discovery but the stakes are also higher. Success provides a new product with new therapy and commercial profit to allow the continuation of the discovery and development cycle. The teams involved in drug development may include whole departments: drug safety (toxicology), drug formulations, process research and clinical pharmacology. The scientist who directs such an inter-dependent and complex organization is typically the Vice President for Research. Thus, the story of how modern medicines are developed is illustrated in the following chapters by responsible management scientists of the pharmaceutical industry.

No attempt has been made to cover all phases of the development process, but rather the authors have sought to depict typical aspects with real and practical problems. These range all the way from chemical process research to modern techniques of formulation and new areas of developmental research that have yet to provide sure evidence that practical new therapy will result.

With the exception of Chapter 2, which was added more recently, the chapters in this book originated from talks that were presented at a symposium for science writers held at Rockefeller University in May, 1972. The authors sought to provide illustrative material to describe the nature of the developmental research that goes on in the pharmaceutical industry when an exciting new discovery is transformed into a useful product. The symposium speakers were introduced by Dr. Thomas Carney and excerpts from his remarks are still pertinent:

"Today we will emphasize work that comes after discovery, the work that make a discovery valuable and useful.

"Any time we talk about a tangible, useful product we talk about something that involved development. Every drug that is made available has involved an extensive development program following the initial discovery of the product. In many cases it is the contribution of the development experts that has been the difference between total failure and the making available of a valuable drug.

"A drug that has an effective activity period of only a few minutes in the body is not of much use. The development of a formulation, or a chemical derivative, or a related compound that will extend the activity for hours, or increase the activity many fold is the difference between success and failure.

"A steroid that costs several thousand dollars per milligram—and many of them have—will not find much use in medicine. The development of a practical method of production that would reduce the cost substantially must be considered as an important contribution.

"A drug that has shown beneficial therapeutic activity is not practical if it is so irritating that it cannot be tolerated by the patient. By the proper manipulation of the pharmaceutical form the harmful side effects are eliminated, and the drug becomes part of the doctor's armamentarium.

"Today, at least in the pharmaceutical industry, development departments must be staffed by people with the same degree of research sophistication as in our exploratory research departments. We must use the most talented physical chemists to interpret results and to help predict means by which drugs are absorbed. Pure mathematicians help develop models for predicting drug systems. The chemist, physiologist, pathologist, toxicologist, clinician—you name it and we have it—are indistinguishable in the quality of their backgrounds and talent from people doing exploratory research. Only the objective of their work is different.

"I sometimes think that the only difference between pure research and development these days is one of geography and of motive. If the university is doing research on the adsorption of a chemical through the intestine, it is the purest of pure research. If we do the same job in our laboratories, it is applied research or development, since we are trying to find how to increase the safety and efficacy of a drug.

No drug would ever reach a patient without a major developmental program between discovery and distribution."

Thus, the chapters in this book will demonstrate the nature of developmental research. It should be emphasized, however, that the industry relies on academic and institutional research to provide much of the new knowledge that makes the applications possible. Industrial scientists maintain close association with their counterparts engaged in academic research. Academic research has provided better methods to synthesize compounds, new techniques to analyze them and improved biological methods to evaluate them. More than that, academic and institutional research has provided new concepts in molecular biology that have changed industrial approaches to new drug discovery and development. Biology is the new frontier of research, but chemistry continues to provide the basis for understanding biology at the molecular level. This book will show how the pharmaceutical industry links the chemist, the biologist and the physician into a cooperative team working with a common goal. Nowhere else but in the pharmaceutical industry is this teamwork so well developed.

No attempt here is made to catalogue or enumerate the many contributions of the pharmaceutical industry that we almost take for granted. For these the reader must look elsewhere. What this book tries to do is to illustrate with real examples the challenges of development research toward new therapy provided by the researchers who are doing the work today. All the papers from the 1972 seminar have been revised and brought up to date. They provide an insight into an important aspect of the way new therapy will be provided in the future to prevent, control or cure the diseases that still afflict mankind.

Suggested Reading

Cavillito, C.J. Interactions of Science, Economics and Politics in Drug Discovery, Development and Delivery. *Drug Development Communications*, 1974-1975, **1**, 259-285.

Clarke, Frank H., Editor. *Annual Reports in Medicinal Chemistry*, Vol. 11, New York: Academic Press, 1976.

Clarke, Frank H., Editor. *How Modern Medicines are Discovered*, Mount Kisco, N.Y.: Futura Publishing Company, Inc., 1973.

Burger, Alfred. Behind the Decline in New Drugs. *Chemical and Engineering News*, September 22, 1975, pp. 37-46.

Hahn, Fred. E. Strategy and Tactics of Chemotherapeutic Drug Development. *Die Naturwissenschaften*, 1975, **62**, 449-458.

CHAPTER 2

The Role of Technology in New Product Development and Quality Assurance

A. C. Shabica

Modern society without medicines would be impossible to imagine. The development of medicines is taken for granted along with their availability, quality, safety and efficacy. The demands of the medical profession and the public's insistence on obtaining quality products may be summed up by the demanding drug regulations that guide the operations of the pharmaceutical industry.

That research and development are required for product quality goes without saying. Which came first is neither certain nor in the scheme of things very important. Rather, what is vital is the relationship of one to the other—so much so that a review of the interrelationship, and the impact of government regulations as a preface to drug development, is essential.

Such a review should give the reader a better understanding of the critical role played by technology in the whole scheme of pharmaceutical new product development and quality assurance.

That present day drugs can prolong man's expectancy of a long and useful life is not questioned. While society has long insisted on quality drugs, the science of quality drug development and preparation has been brought to the forefront only in the past 15 years. Indeed, pharmaceutical preparations have evolved from an art, as practiced no more than two generations ago, to the scientific, automated techniques and quality assurance developed and practiced by industry and government today.

How did this come about? Was it because of the 1938 drug

law with its subsequent changes in 1962 and 1969, or has it been a more recent change in the attitudes of the teaching profession, industry and government?

The record speaks for itself. It was 156 years ago, in 1820, that a *voluntary* physician-pharmacist effort resulted in the Pharmacopeia of the United States. As far as we know, this was the first attempt to standardize pharmaceutical preparations *voluntarily*.

This was followed by the founding in 1821 of the Philadelphia College of Pharmacy, the nation's first. The leadership of this school in controlling drug quality also is a matter of record.

It remained for the latter stages of the industrial revolution, with its scientific and technological advances, to lay the foundation for a modern pharmaceutical industry.

The invention and use of machinery shifted the preparation of pharmaceuticals away from the pharmacist's shop to the manufacturing plant. Techniques and instruments were developed which permitted, for the first time, reproducible products and accurate testing of their quality.

One cannot say, then, that in the United States, where there were no laws requiring drug labeling *and* no restrictions on ingredients or therapeutic claims, that the movement for drug standardization and quality manufacture in the 1800's was led by individuals and private organizations.

This was a step forward, but as it turned out, only a first step. The major breakthrough came more than one hundred years later when Congress passed the 1938 Food & Drug Act and thereby recognized the public demand for safe drugs. Although it may be only coincidence, one could say these laws were responsible, albeit indirectly, for stimulating pharmaceutical research and development.

In other words, the insistence on a major role for research and development was recognized and articulated by government and industry. The role of technology in assuring the public that drugs must be of high quality by a process starting with research through development, production, and final use, was inevitable.

Since the 1938 regulations, the claim has been made many times that all the major pharmaceutical companies are good and not one in particular outranks the rest; further, that all the major companies have excellent research and development programs, and that most problems arising in the industry apparently come from the small firms that do not have the necessary resources to foresee or prevent problems.

While this may have been true to some extent years ago, these statements no longer stand inspection, especially if one examines the records of product recalls and the demonstrated performance of both large, medium and small companies.

Even large companies have been faced with critical problems not because of oversight, negligence or lack of effort but only because the development methodology available even 10 years ago could not establish the absolute potency and the more subtle toxicities of all marketed products—as was so dramatically demonstrated by the thalidomide incident. Even though this was an isolated incident, it has been referred to repeatedly as a model for the general performance of the industry.

Despite all the finger pointing, the thalidomide tragedy is not at all a typical case. The major companies have always depended heavily on very carefully planned and executed research and development. This, coupled with good manufacturing and quality control, has provided high quality products which in turn established and maintained their reputations. However, had it been left at that the industry might have been in for even greater surprises and difficulties than it experienced in the 60s. But modern pharmaceutical thinking was shaped *prior* to the 1962 regulations.

It was clear, even in 1938, and certainly by the end of World War II, that the old way of developing products no longer satisfied the needs of a modern pharmaceutical company, the people it served or, as it turned out later, the regulations that would finally govern it.

It became equally clear that any program was doomed to failure if it stressed rapid product development and routine production while ignoring the responsibilities and commitment of all personnel involved in a product—beginning with the researcher himself. Such programs would have been catastrophic for the industry if they had become common practice. Fortunately, it was uncommon to expect companies to meet preconceived time schedules and sets of specifications without looking carefully behind those specifications.

For the most part the industry recognized the broader responsibilities placed on developmental research groups, and especially on all employees. By 1965, pharmaceutical scientists and technologists began to hear such terms as *zero defects, product security* and *pride in quality.*

With this recognition came a shift of emphasis from trying to guarantee quality through specifications to high quality product development, and the added quality assurance programs. In other words, new dimensions were willingly added to guarantee product quality and satisfy the most demanding of the industry's critics.

Needless to say, if many companies had not reached the degree of sophistication in R&D prior to 1969, product failures would today be more frequent than they are—production, quality control and quality assurance, notwithstanding.

Lest one be persuaded that quality assurance is just another term, let me quote the following definition as it has been reported in today's legal, regulatory and medical climate:

> TO ASSURE THE PROFESSIONAL USER OR ULTIMATE CONSUMER THAT EVERY LOT OF A PRODUCT IS PRODUCED UNDER THE BEST KNOWN METHODS AVAILABLE TO THE INDUSTRY AND THAT IT CONFORMS TO SPECIFICATIONS AND SAFETY REQUIREMENTS AND THAT EACH DOSE OF EVERY PACKAGE DISTRIBUTED WILL FULFILL THE LABELLING RELATIVE TO EFFICACY AND TOXICITY . . . AND THAT AT ANY TIME, THE HISTORY OF EVERY PACKAGE DISTRIBUTED CAN BE TRACED BACKWARDS THROUGH THE WHOLE PROCESS . . . AND THAT EVERY LOT OF A PRODUCT WILL BE TRACEABLE FORWARD UNTIL IT REACHES WHOLESALERS OR MAIN DISTRIBUTION CENTERS.

(From *The Endless Struggle for Quality: Drug Manufacturing from the Sumarians to Current Practices of the World-Wide Pharmaceutical In-*

dustry, by Charlotte D. Duff, 1971. Prepared at the request of the Pharmaceutical Manufacturers Association International Section.)

An official but complicated definition, to be sure. Later, a simpler and more practical definition will be presented.

To return to the regulatory evolution and its impact on product development, one is easily convinced that as the '60s came to a close the pharmaceutical industry had reached new levels in controlling quality. To the credit of a number of companies, quality was no longer "inspected" into a product; it was *built in* from initial product research to ultimate distribution. The better the quality of the development effort, the fewer the problems encountered by production, quality control and, ultimately, the patient.

What are some of the changes in new product development that took place in the '60s which now assure us of quality products and a minimum of difficulty?

To answer this, one must first look to regulations as they existed at the time.

In late summer 1969, the Food & Drug Administration published the first revision of *Current Good Manufacturing Practices* since adoption of the regulations in 1963.

Although at first glance the 1969 version did not appear different, a detailed inspection showed that additions had been made that clearly spelled out the intent of the 1962 law. More importantly, they described a major change which was, and still is, considered revolutionary in its concept and impact.

Specifically, "A product is considered impure if production methods, facilities and controls do not conform to current good manufacturing practices (CGMP) requirements for safety, potency, identity, quality and purity."

Simply stated, a product will be considered adulterated if it is not produced under conditions accepted as current GMP, even though the product meets all specifications of potency, identity and safety. Clearly this now placed a new and additional responsibility on all development areas, as well as on production.

While the intent of the law was clear, i.e., to protect the consumer from poor manufacturing practices initiated at the research and development stages, the lack of detailed guidelines makes the manufacturer's task in achieving regulatory compliance exceedingly difficult.

To cope with this problem, pharmaceutical R&D management has been in the forefront of product development and technology by going beyond the minimum requirements of the regulations. This means being innovators and not just followers.

To accomplish this laudable goal, it is now expected that product development and quality assurance will begin as early as the isolation or synthesis of a new drug substance. Following this, it will motivate all scientific personnel during the development stages of a product to monitor production, storage and distribution and, finally, assist production and control in maintaining high standards of performance, as long as the product is in use.

To better understand the role of chemical development and its interdigitation with other disciplines, let's examine the four separate but interdependent areas of product development: clinical research and development, chemical and chemical engineering development, pharmaceutical research and product development, and package research and development.

During each of these phases, several objectives arc sought that are closely watched and accurately defined so that the integrity of the product can be guaranteed, especially as to reproducibility, efficacy, safety and the good manufacturing practices regulations.

Clinical Development

To achieve new product quality assurance, the objectives of clinical investigation include detailed definitions of the product as related to use, which become a part of the literature and labeling. This includes details on action of the product, indications, contraindications, warnings, precautions, adverse reactions and dosage and administration. This phase of the development of modern medicines will be discussed in detail in Chapter 7.

Chemical and Chemical Engineering Development

Successful companies founded in the 30s or late 40s recognized research and development as an indispensable and integral part of new product development, especially as it related to the preparation of active pharmaceutical ingredients and formulated products. In order to prepare for expected product demands and to meet inevitable competition, these same companies either started R&D departments or expanded existing ones.

Of special interest was the emphasis on newly emerging technology as embodied in the disciplines of chemical development and chemical engineering. The addition of this technological arm in the pharmaceutical industry made it possible to move rapidly to improve existing product lines and to develop new ones. Major examples of such successes are the cortisone synthesis and early antibiotic development—both striking technological achievements for that period. Unfortunately, the industry could not get across a convincing enough message of accomplishment to the general public.

Because the public did not understand or appreciate its major achievements, the industry found itself in the '60s subject to the government regulations already mentioned. Meanwhile, a concerned public had come to demand an accounting for all new developments. It is these pressures that have brought about a very profound change in the approach to process R&D.

At the risk of oversimplification, one might say that chemical development progressed from what could be called a "bread and butter" approach to process development in the '40s and '50s to the theoretical approaches embodied in reaction engineering and computer technology in the '60s, then back to the "bread and butter" approaches in the '70s—thus coming full circle. The changes in the late '60s and early '70s emerged from legal and social pressures brought to bear on all industry, not just the drug firms.

To those pioneers in the field of chemical technology it would appear that the developments through the '40s and '50s came about in an era of largely favorable publicity, and the negative view of industry that followed later was to be expected. But not expected was the extent of the criticism.

If we look back to the "old days," we find that the then relatively new technique of process research and development had four major objectives:

1. To operate a pilot plant wherein material was produced for clinical trials, advanced biological-toxicological testing and, last but not least, pharmaceutical R&D studies.

2. Technologically to choose and develop a chemical process for a product provided by the research chemists and to establish operating yields and quality.

3. To obtain the operating know-how to demonstrate a full-scale chemical process.

4. To construct and transfer successfully a pilot process to a large-scale plant, whether it be a batch or continuous process.

Today, these objectives have not changed qualitatively; however, in-depth studies and extensive additions of new techniques have given the modern process development group a new look.

A modern chemical development group should be made up of a scientific and technical staff organized into at least two sections, process research and process development—the former being made up primarily of chemists and the latter of chemical engineers. The process research group is responsible for the evaluation of alternate chemical syntheses and chemical optimization of the process of choice. Process development employs chemical engineering technology to scale up the laboratory process to manufacturing equipment. In addition, it must design, equip and start up operation of the plant at any chosen scale.

Interface with other R&D groups such as biological research, chemical research, analytical research and development, pharmacy research and development, research and production is maintained throughout the course of chemical development. Interdigitation of these groups is discussed briefly later on.

Although there are a number of working concepts and philosophies for conducting chemical development, two of the more common ones involve compartmentalization, where the chemist completes this process research work and then turns it over to the chemical engineer who continues with the technological development phase. The results of the chemical engineer's studies are then transferred to engineering for final design and plant construction and finally to production for operation.

Success of this approach is dependent, of course, on close collaboration and communication at each transferral stage. An engineer is assigned to the project late in the process research stage. Similarly, plant engineers and production people are brought in during the final process and development stage. This system permits engineering and production input when and where it can count before a project is finalized.

An alternate working concept is the project team approach headed by a chemist or a chemical engineer. The team is formed at the beginning of the project, and it involves mutual and reciprocal consultation and observation of activities in the lab by the engineer and in the pilot plant by the chemist, with emphasis shifting gradually from the laboratory to the pilot plant as the project matures.

The major advantage of this approach is that it takes maximum advantage of the specialized training and skills of chemists on the one hand and chemical engineers on the other, while at the same time insuring a smooth transition from the essentially chemical-scientific approaches of the laboratory to the basically engineering-practical approach in the pilot and production plants.

In other words, throughout the development process and into production, a chemical engineer deals with the chemist as a partner, thus insuring the achievement of economic goals with a reasonable degree of safety. The later stage, i.e., the design, construction and operation of larger scale pilot plants, semicommercial units and the finished plant itself, requires very little science and chemistry but more mechanical skills and aptitudes. At this stage, the chemical engineer develops a liaison with the engineering and production departments and is permitted to go his own way. The chemist may be brought back into the picture only when plant start-up troubles develop, if they do.

In the '50s and '60s, the application of the prepilot and pilot plant studies to a practical commercial process and to plant design was the primary objective of the process and chemical engineering section. It required considerable practical experience in process engineering. The rewards were mostly economic in that they lessened the investment risk of new processes by making sure that reproducibility of quality and yield under safe conditions were pretty much guaranteed. These rewards satisfied the most demanding of administrations and at the same time gave great personal satisfaction to the chemists-engineers involved.

As has been learned by experience, many chemical reactions that work well in small volumes (one quart to one gallon) perform poorly when expanded one hundredfold or more at the pilot plant level. Differences in surface to volume ratio, rates of heating and cooling, degrees of mixing and rates of addition of reagents are only some of the variables that profoundly affect yield and quality and thus the economics of the process.

To determine the correct dimensions of these parameters often was a tedious job, for it was mainly trial by error. More recently this approach has been simplified by applying mathematical modelling techniques and finally computer simulation.

Once the major variables have been pinpointed and their limitations established, the validity of a process for large-scale operation is finalized. Other dividends of the study also may be achieved. These may be considered as beneficial by-products of the pre- and pilot-plant stages; i.e., sufficient material being produced to explore the full potentiality of the new product in its field of application. In addition, operating know-how is provided that is of inestimable value in starting the full-scale plant.

In the past, a development chemist was not limited in his choice of chemical reactions or operating conditions as long as he could show satisfactory yields, economy of manufacture, safety and quality of end products.

With inexpensive labor, cheap raw materials and inexpensive equipment available, and with relatively short delivery times, the chemist and chemical engineer had *n*th degree freedom to innovate—usually quality and safety being their only limiting factors.

In more recent times, scarcity and high cost of skilled labor, scarce and increasingly more expensive raw materials with shortages of equipment and, equally important, long delivery times have led to greater restrictions on the chemistry available to the chemist-chemical engineer for translation to the pilot plant and finally the finished plant.

Although the importance of chemistry has not lessened with these limitatations, the close collaboration of the chemist and chemical engineer surprisingly became a secondary issue. The era of continuous operations began with automation. Computerization and application of kinetics and thermodynamics to process optimization and reaction engineering became familiar words in the development lexicon. As a consequence, chemistry took a back seat. With these heralded changes, engineering attained a long-awaited stature of its own. In other words, engineering know-how began to dictate to chemistry, rather than the other way around.

This would have continued except for government regulations now extended from the FDA to OSHA (Occupational Safety & Health Act), environmental restrictions at the local, state and federal level, disappearance of certain raw materials either through government regulation of diminishing world supply (petrochemicals, energy shortages, equipment shortages) and, finally, a more questioning consumer public.

With these handicaps, optimization and exact reactor design has almost become a secondary matter. This is not to say that the theoretical approaches developed in the last 10-15 years are obsolete and should be thrown out—not at all. However, they are kept in perspective with more mundane chemical limitations and pressing new requirements.

This brings us back full cycle to the combined chemist-chemical engineer approach.

This partnership relies heavily on the chemist once again and the combination is being strengthened as time goes along in order to comply with social, legal and marketing pressures.

Some of the problems already mentioned have been increasing over the past few years. In order to see clearly the evolution of a technical process and to observe how these problems affect the approach to chemical development, let's look at a typical example involving the overall development of an active drug from chemical research conception to chemical manufacturing.

Although the following discussion refers to multistep, totally synthetic processes, it should be noted that many pharmaceutical agents, such as the steroids and antibiotics, are obtained directly or by chemical modification products derived from natural sources.

In these cases, the development team concerns itself with processes such as extraction, fermentation and separation rather than chemical transformations. The technique and planning, however, are subject to the same kind of detailed study and analysis as outlined for purely synthetic substances.

Initially, a potential drug substance is produced in relatively small quantities for beginning efficacy and animal toxicity studies. The more elaborate biological, pharmaceutical formulation and clinical studies in man require very large quantities. More often than not these quantities are in the hundred- and thousand-kilogram range.

Further, it is essential that these substances are prepared in a reproducible manner and in a high state of purity at a price that the consumer eventually can afford. Simply stated, the role of the development chemist and chemical engineer is to develop a chemical process for the large-scale preparation of drug substances, to provide sufficient quantities of the drug itself to support the other R&D groups involved in drug development and, finally, to provide the means for ultimate manufacture.

In approaching his responsibility, the development chemist finds his efforts limited by two constraints, *time* and *processing*. To appreciate the time constraints, it is necessary to recall the major decision points in the development of a new drug:

1. The decision to obtain an IND (investigational new drug application) and conduct initial safety and efficacy studies in man.

2. The decision to conduct extensive trials in man.

3. The decision to apply for an NDA (new drug application) in order to market the drug.

To Point 1, the requests for drug quantity are small (usually less than one kilogram) and these supplies generally are produced in the research laboratory by the chemist who first synthesized the drug candidate. The compound may, in fact, be just one of a great many undergoing preliminary biological screening. If selected as a potential clinical candidate, kilogram quantities will be required for initial formulation and extended toxicity studies in support of an IND. Such quantities, generally, are too large to be carried out in the laboratory and for safety reasons as well as time are produced in the pilot plant. It is at this point that the development chemist enters the picture.

From this start and through Point 2, the main work of the development group is carried out. At the point of extended clinical trials, the drug should be made by the final manufacturing process so that data gathered and evaluated in these trials eventually will reflect actual experiences with the finally marketed product. After Point 3, only minor modifications of the process are permitted; the basic chemistry, however, remains unchanged.

Processing criteria, which enter the development chemist-chemical engineer analysis of potential synthetic schemes are: product quality, safety, environmental impact, process operating conditions, equipment requirements, product yield and cost.

These factors have not been listed in their order of importance, except for the matter of product quality. This always remains first and foremost in the development scheme. For the other criteria, however, the relative attention devoted to each will vary with the particular step in the synthesis and, more often than not, with the general evolution of the process itself. This is necessarily so, as these factors are not independently variable.

For example, in labor intensive processes, the choice of large equipment to achieve high volume may be exercised even though lower product yields may result. With low-cost raw materials, the contribution to final cost is not as great. Conversely, with raw material intensive processes, small equipment may be chosen because labor costs tend to be a minor factor, and further, small batches can minimize loss of expensive materials. The development team of chemists and chemical engineers must establish the optimum balance among all of these considerations; however, in no case is quality compromised. A closer examination of these criteria provides insight into the development process.

Safety

The process must be fully defined and predictable so that it can be carried out routinely by technically skilled operators who have a minimum of formal chemistry training. Especially hazardous chemicals should be eliminated wherever possible and, where not, the process handling designed so there is no direct exposure. Extreme reaction conditions, particularly very high- or very low-reaction temperatures, are avoided when possible as this increases the possibility of equipment or human failure with a concurrent increase in the chance of "something going very wrong."

While desirable, it is, of course, not possible to carry out the whole of chemistry in neutral aqueous solutions at room temperatures. Therefore, real and potential safety problems arising from exotic solvents and operating conditions must be recognized and dealt with in both process and equipment design.

Especially hazardous reactions can be carried out in specialized and dedicated equipment, located in protected manufacturing areas. In this regard, the relative hazard of each step of the synthesis is assessed and the equipment designed in such a manner as to give maximum protection to the operator. Loss of equipment is tolerable, but not loss of human life.

Environmental Impact

The consequences of a chemical reaction on the environment has become increasingly important as more stringent legislation severely limits the effluents that can be discharged. While it is possible to design scrubbing units to protect reactors from accidental discharge of gases and to treat reaction sidestreams chemically, doing so does not guarantee elimination of accidental spills and gas discharges. Mechanical failure of such auxiliary equipment does occur from time to time. Because of this, it falls to the development chemist to devise schemes that have a minimum of side-products, which must either be removed or recovered from the reaction process.

As more information becomes available on the toxicity of such simple chemicals as ammonia and zinc in water effluent, more effort is directed toward eliminating these chemicals either from the synthesis, or by highly sophisticated and expensive recovery systems.

Process Operating Conditions

The more general term, *process operating conditions*, not only encompasses aspects of safety, environmental impact and equipment need but also focuses on such factors as mechanical properties of the reaction mixture (flow rates through the reactor openings, stirring characteristics, separation rates of different reaction phases, filtration properties, etc.), as well as the operational restraints that might pertain. Under normal circumstances, it is not possible to put large reactors through rapid temperature cycles, nor is it normally possible to charge reagent or quench reactions rapidly. Such restraints require special reaction design, such as can be achieved by flow-systems.

On the other hand, it is feasible to carry out certain functions more easily in large-scale equipment than in the laboratory. For example, multiple extractions and filtrations at elevated temperatures are routine on a large scale but very awkward in the laboratory setting.

Equipment Requirements

The equipment requirements for a synthetic process are dictated by the process operating conditions. Large-scale manufacturing equipment, even pilot plant equipment, is exceedingly expensive and demands expensive support capitalization, so that it is desirable to design processes for either simple, dedicated plants or already existing multipurpose equipment.

In the case of specialized processes such as fermentations

which are highly sensitive to cross-contamination, or of reactions, which could introduce contamination (such as mercury or lead) in subsequent processes, or for very hazardous work such as dissolving metal or high-pressure reactions, dedicated equipment is installed. Development must be highly committed to such processes, and every alternative explored in order to minimize major capital expenditures.

Cost

While process capital equipment can contribute heavily to initial or "start-up" costs of synthetic process, in the long run a more significant factor is the expense of the process itself. Thus the cost of raw materials, the time and labor of the process, recovery of raw materials, waste stream treatment and energy requirements all enter into the final price of the drug. It is one of the challenges to the chemist-engineer team to develop a process that reduces costs.

Product Quality

The standards required of pharmaceutical agents are such that it is not only necessary initially to produce the purest product possible but also to guarantee that no new impurities are introduced during the lifetime of the drug. The definitive toxicology, bioavailability and pharmacology originally carried out on the drug reflect the absolute purity of the product, and this purity must be maintained from batch to batch to allow for predictable prescribing. If a choice must be made, it is usually better to have a lower yield of higher quality product. Where it is not possible to remove completely any impurity, every effort is made to characterize its toxicity and biological effects and to control its content in the finished product.

In order to appreciate the care given to the absolute control of the finished product during clinical trials and throughout its lifetime, the efforts of the quality assurance group become paramount in monitoring all phases of research and development as well as production operations. This is done to insure that good manufacturing practices are met and that detailed identity and

Figure 1. Large scale equipment is operated by chemical engineers.

history of the product and its raw materials are documented prior to release, either for clinical studies or commercial use.

Similarly, the quality control laboratories continually check the receipt of all raw materials throughout the development and production of the the active ingredient and final pharmaceutical formulation and packaging.

Each raw material or intermediate used in the synthesis is quarantined on arrival and is not released until quality control or analytical R&D has determined its acceptability by inspection and laboratory testing. A typical example of the kinds of checks and the number of check operations for each raw material is listed below. Depending on the nature of the raw material and its stage of use in the synthesis, the kinds of checks and number of check operations vary.

Receiving Clerk Checks	*Kind of Check*	*Number of Check Operations*
Supplier	1	1
Purchase Order Number	1	1
Title and Product Code Number	2	2
Supplier's Lot Number	1	1
Container Count	1	1
Recorded Weights	1	1
Container Condition	1	1
Specific Storage Requirements	1	1
	9	9

Quality Control Sample Clerk Inspection		
Confirm CIBA-GEIGY title	1	120
Confirm Product Code	1	120
Confirm Supplier's Lot Number	1	120
Confirm CIBA-GEIGY Purchase Order No.	1	120
Containers opened and examined	2	24
Sample Taken	1	12
Sample Drums Initiated	1	12
Sample Blended and Labelled	1	1
	9	529

Laboratory Analysis of Sample

Log Sample and Assign Quality Control No.	1	1
Chemical Identity	1	1
Chemical Analysis	10	10
Ignition Residue	1	1
Heavy Metals	1	1
Clarity/Color	1	1
Water Content	1	1
Writes Report	1	1
Checked by Supervisor	1	1
Confirmed by Manager	1	1
Release Issued	1	1
Preparation of Labels	1	120
Label correctness checked	1	120
Sample Clerk affixes labels	1	120
Future Reference Sample filed	1	1
	24	381

Product Yield

Because most processes are multistep, the effect of each step's yield is geometric on the cost of the final product. For this reason, much effort is directed toward improving the productivity of each step in the sequence. This can be done by increasing the chemical yield of the reaction and by decreasing operational volumes and processing times. Usually, final manufacturing productivity can be enhanced over that observed in the laboratory by evolutionary operation (effect of minor changes of operating variables) during plant operations. Such "fine-tuning" of the operating variables is primarily handled by the chemical engineer with concurrent laboratory support from the chemist.

In determining his strategy, the development chemist initially reviews the existing chemistry from the standpoint of safety, equipment need and process operating conditions. With the team engineer, this process may be run in the pilot plant; material produced by this trial provides for indepth toxicity studies and initial clinical trials in man. The processing information gained, coupled with concurrent laboratory studies, furnishes the basis for targeting the best possible processing chemistry, be it the original process created by the research chemist or a new synthesis that appears more feasible from a large-scale viewpoint. This focus on the best *possible* processing chemistry is important, since the manufacturing process should undergo only a minimum of change throughout the lifetime of the drug.

Because of the large number of variables to be explored, and their interrelations, a heavy reliance is placed on the chemist's synthetic skills and his knowledge of manufacturing possibilities. (The fact that manufacturing reactors are opaque, bolted to the floor and behave like Dewar flasks alters the perspectives of a particular synthetic scheme. A one- or two-degree exotherm observed in the relatively small laboratory experiments can become catastrophic in 2,000 gallon reactors unless prior provision is made for adequate cooling. The reaction kinetics must be well-determined to allow nonchemists to run the reactions reproducibly, and the reaction mixtures must be sufficiently fluid to be pumped through transfer lines.) Based on this knowledge, alternate synthetic routes can be evaluated by inspection for the criteria of safety, environmental impact, process operating conditions and equipment requirements. Alternates that satisfy these criteria are then evaluated for competitive cost by compilation estimation techniques.

This evaluation process limits the possible synthetic routes severely; those remaining must be studied experimentally for yield and product quality. Even here, however, the classical techniques of evaluating each operational variable may not be necessary.

By mathematical modeling of the process, it is possible to carry out a limited but well-planned series of experiments that will provide the data needed to determine the important variables that affect product yield and quality. For each simple process, simple models are generated from which optimum conditions may be estimated by inspection. More complex processes, particularly those involving competing reactions, require more precise models and more refined data such as that derived from kinetic studies.

It is often observed that simple, well-established chemistry requires developmental modification prior to its consideration as a manufacturing process. Initially this may be directed at cost reduction by shortening the synthetic sequence. A more careful analysis can actually lengthen the step sequence due to such factors as operational safety and handling characteristics, environmental or equipment considerations. The following example shows that these can balance for a net gain in production efficiency.

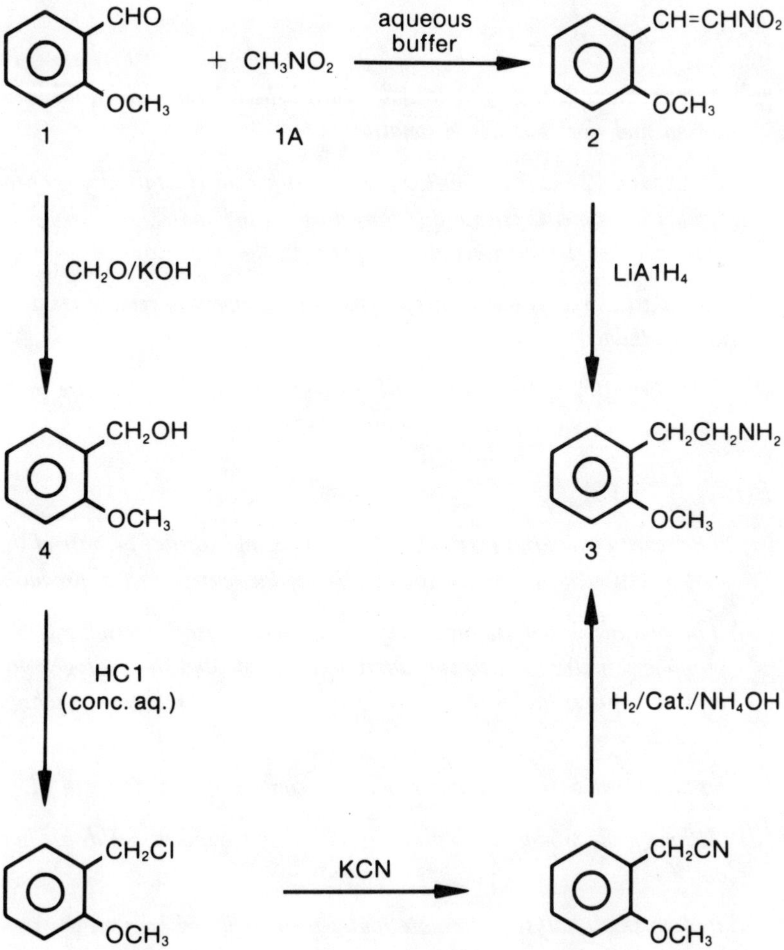

Reaction Scheme

Compound 3 (Scheme) was required as an intermediate in the synthesis of a drug candidate for indepth clinical trials. The sequence (1) — (2) — (3) had been utilized in the research laboratories for rapid access to (3) and was successfully carried through the pilot plant. Based on this experience, the following critique was developed:

FOR (1) + (1A) → (2)

(a) Nitromethane is highly toxic and requires special pumps to avoid cavitation and thus possible detonation.

(b) Compound (2) is a low-melting solid with poor crystallinity properties; it is a potent vesicant and thus difficult to handle, and it requires extensive low-temperature drying prior to the next stage.

(c) The waste streams would require special treatment to remove residual nitromethane.

(d) The chemistry is traditional and well-known; the reagents are inexpensive, and yields are acceptable (86-89%).

FOR (2) → (3)

(a) The reaction requires excess lithium aluminum hydride; because of the cost of this reagent, it contributes 36% of the overall cost of product.

(b) The volatility and flammability of the tetrahydrofuran solvent, the pyrophoric nature of lithium aluminum hydride and the hydrogen gas evolved requires that this reaction be run in special equipment in a high hazards area.

(c) Salts formed as side products must be contained and disposed of.

(d) Although the product is obtained in relatively pure form, yields are low (50%).

Based on this analysis, alternate routes were explored both on theoretical grounds and through reduction to practice in the laboratory. The most satisfactory is given by (1) → (4) → (5) → (6) → (3) as shown in the Scheme. It should be noted that this contains twice as many chamical stages as the original synthesis; however, overall processing *time for the two se-*

quences is virtually identical. Cost analysis revealed the longer sequence to be the least expensive due to lower costs of reagents and solvents. The same type of critical analysis can be applied to this sequence:

FOR (1) → (4)

(a) *Formaldehyde is toxic; waste streams must be treated for its removal.*

(b) *The reagents are inexpensive and available; the reaction is rapid.*

(c) *The reaction product (4) is not isolated but carried in solution to the next step.*

FOR (4) → (5)

(a) *Hydrochloric acid is available and inexpensive; although it is caustic, its waste streams are easily neutralized for discharge.*

(b) *The reaction is rapid and quantitative; the product is not isolated but carried in solution to the next step.*

FOR (5) → (6)

(a) *Potassium cyanide is highly toxic, so the reactor must be protected against accidental venting of hydrogen cyanide gas. Waste streams must be either oxidized or remaining cyanide removed before discharge.*

(b) *The reaction product (6) is crystalline and easily purified.*

(c) *The overall yield for (1) →→ (6) is greater than 90%.*

FOR (6) → (3)

(a) *The low-pressure hydrogenation must be carried out in special equipment in a hazards area.*

(b) *The initial stock of catalyst requires a large capital investment; however, it may be reused and recycled.*

(c) *The reaction is rapid and clean, product isolation is simple and waste streams require only neutralization.*

(d) *The yield of product is greater than 90% and the quality of product is at least as good as that of the shorter sequence.*

Thus on balance, the objections raised by the longer process are considered minimal with respect to the alternatives. This, coupled with identical processing time and lower cost, make it an alternative process.

This type of analysis is carried on continually in the development of manufacturing processes for new drug substances. Most often the chemistry involved is not so straightforward, nor are the molecules so simple. The objectives are, nevertheless, the same: to take molecules that have been little more than laboratory curiosities, explore and understand fully their chemistry and the possibilities for their synthesis, and finally arrive at methodology that allows their safe, consistent and economical preparation.

In all of the discussion above, it is difficult, if not impossible, to define exactly the chemist's role as distinct from the chemical engineer's. In the final development of a process this distinction is unimportant. What is important, however, is to distinguish the chemically oriented problems from the mechanical or equipment problems. Only where this differentiation is clear can one assign a direct role.

Obviously, close collaboration throughout the project between the process research chemist and process development engineer is essential to success. Equally important is close collaboration between the process engineer and production chemist during the final development stage. Presuming the above work to have been done thoroughly, pilot-plant demonstration and conceptual plant design is effected—the latter being turned over to the engineering department for construction design and installation.

In spite of the best laid plans and the quality of the development work, plant difficulties arise from time to time for reasons beyond the control of production people. These difficulties may be related to changes of raw materials, undetected changes in process conditions or substitution of highly trained with less trained personnel.

These problems may relate to yield or then again to purity. In both cases, solution of the problem becomes urgent and if not discernable immediately by production, development personnel are called in to troubleshoot. By and large, it is expected that these problems will be kept to a minimum, and, in effect, the ease of solution measures the thoroughness of the previously completed development program.

Presuming all has worked out well, the active ingredient is subjected to the same systematic studies by pharmacy research and development during formulation studies and is carried on to production in pretty much the same fashion.

Looking to the future, successful chemist-chemical engineer teams today must know more about the total R&D process (biological, pharmaceutical, medical) as well as the "state of the art" technology than their predecessors. They will find solutions to more difficult environmental and energy problems, even with engineering principles restricted by marketing, legal, patent and government regulations. Opinion, or personal preference, will no longer be accepted in the day-to-day problem-solving arena. In summary, greater and greater demands will be made on the development team and unless solutions are found, new product development will decline.

Every chemist-chemical engineer team must find new ways to do more complex things in even shorter periods of time. Forward looking companies will plan for as many degrees of freedom as possible, in their chemical operations, in order to assure innovative progress.

Pharmaceutical Product Development

Looking toward pharmaceutical product development, we find the objectives not much different than they were a generation ago. However, as in the case of chemical development, the new tools and methodology developed over the last twenty years have had a major effect on the approach to these objectives.

To compare today's approach to pharmaceutical design to that practiced 20 years ago, is apt to lead to the conclusion that the pharmaceutical industry was amazingly lucky in producing acceptable products with only a minimum of information.

This, in fact, is not true. Quite apart from efficacy and safety, which were primary requisites for responsible companies, the definition of *acceptable* was based on different criteria from that used today. Assuming, *a priori*, a medically accepted product, one of the yardsticks for acceptability a generation ago was the cost of making a pharmaceutically elegant product. Could the active ingredient be made economically? Did the packaged product appear visually acceptable? Did it taste good? Did the tablets have the proper sheen? Was the color appealing to the eye? Did the tablet crumble or maintain its integrity? These were the first considerations of the formulator at the time.

Probably the second criteria for an acceptable product concerned stability. In spite of a limited capability, only a relatively few of the products that were considered stable in the past have been proven with modern scientific tools to be unstable.

Probably the major factor that minimized those problems was the fact that the pharmacist was dealing with chemically inert active ingredients and solid dosage forms (tablets and capsules). For the most part these are more stable than solution systems. A third criteria was absorption of the product itself. If absorption of the drug from the dosage form was considered at all, it was probably thought to be the disintegration rate that was set at some arbitrary standard that could be manufactured, rather than one based on an established biological relationship.

In short, the industry and regulations at the time, for want of better guidelines, set product specifications on either visual parameters, or based them on clinical experience with the routinely manufactured product. The current approach is to determine those relationships that will permit well-defined and scientifically established specifications. Toward this end, a great deal of effort in recent years has focused on redefining specifications or reformulating established products to meet scientific requirements.

Admittedly, there are those who state that many of the new techniques represent, to some extent, window dressing and are not quite as informative as their proponents claim. For the moment, insufficient evidence is available to draw any final conclusions.

However, one thing is certain: 10-20 years ago such terms as *dissolution rates, bioavailability, biopharmaceutics, pharmacokinetics, drug metabolism, blood level studies, indepth stability* and others were not a part of our vocabulary; or if they were, they were only vague terms reserved for the "ivory tower" scientist. Today we wouldn't think of proceeding with a new product or continuing an old product without an indepth study in each of these areas.

For example, in the early "preformulation stage" of dosage form development, *in vitro* dissolution studies are generated in order to gather information about the solubility rate of the drug. Such studies are continued all through the development of the active ingredient and dosage form until the final product is prepared. The dissolution rate is then monitored as part of the regular stability program to assure us that no potential change in biological availability of the drug in the dosage form can occur over periods of time or under adverse storage conditions.

As a result of such studies, many reformulations may be required during the development stage. During these stages no product goes into the clinic for evaluation, even if the dosage form is not the expected final one, without the *in vitro* dissolution pattern conforming to a set of high standards.

We are obviously trying to avoid having a drug appear to be ineffective in the clinic, not due to a lack of efficacy, but because of a lack of uniform preparation or of biological availability.

In addition to dissolution, many development laboratories have as part of their preformulation screening program a partially *in vivo* evaluation of the drug based on transport across an everted rat gut membrane. This alerts the formulator to cases where drug absorption is limited not by dissolution rate but by absorption rate.

Such information tells the formulator that he should not waste his research efforts trying to develop a more rapidly dissoluting form of a drug since its rate is determined by absorption kinetics rather than by dissolution kinetics. In addition, it can also alert the medicinal chemist to consider another derivative that would be more rapidly absorbed and possibly result in superior efficacy.

While one may disagree on the absolute value of such information, we certainly can agree that it does contribute immeasurably to our confidence in the efficacy and safety of the product and, equally important, our ability to reproduce it routinely at the production level.

Although the relationship between dosage form and bioavailability has been recognized for at least 10 years, the application of this relationship to efficacy and toxicity in other test objects has come to the forefront only during the past couple of years.

This is an area in which pharmaceutical development can and has made an important contribution in guaranteeing, if at all possible, the quality of our drugs. If the dosage form of a drug can influence its bioavailability in man, the same would be true for other test objects. Thus, in our screening for efficacy of a drug in other systems, the question arises as to whether an effective drug will be missed because it is being administered in a poor delivery system. This is also true with toxicity studies. If the delivery system is a poor one, the drug may appear to possess low toxicity when, in fact, the human toxicity in a good dosage form might be considerably higher.

When one examines the systems used by biological groups to evaluate drugs, the variations in these systems and the possibility for poorly absorbed drugs become apparent.

Based on the formulator's experience in developing dosage forms with good bioavailability, the biologist employs input from chemistry and pharmacy research and development in an effort to develop general approaches to the evaluation of drugs in all test systems.

It is apparent, then, that those individuals engaged in dosage-form design must be able to handle many areas of expertise. No longer is an ability to operate a tablet press or an encapsulation unit sufficient background for a man to call himself a formulator.

A thorough knowledge of and interdigitation with disciplines other than pharmacy is required. Today's drug researcher, be he chemist, biologist or engineer, is well versed in the areas of physical chemistry, including chemical kinetics, colloid chemistry, thermodynamics and phase rule.

In addition, his biological understanding must be reasonably complete in order to concern himself with designing suitable dosage forms that are well-absorbed, or whose absorption is controlled in certain ways to produce very specific blood level patterns and concomitant efficacy.

Obviously, pharmacokinetics and biopharmaceutics are major fields for today's formulator. A knowledge of chemistry, analytical chemistry, statistics and engineering has become a necessity.

With all of these new tools, the product development chemist finds himself in today's industrial organization with additional capabilities. These in turn have expanded his objectives, which if not achieved will fail to yield the quality of product and reproducibility of production required to avoid product recall. However, when done properly, they contribute to our confidence in the integrity of our products.

A more detailed discussion of bioavailability and the design of drug delivery systems is provided in Chapter 6.

The objective of the pharmaceutical research and development division, then, is not different from the chemical development groups; i.e., to design effective, economic, safe and reproducible manufacturing procedures for the final formulated product.

For this purpose, extensive pharmaceutical laboratories, as well as pilot plants, are maintained in order to carry out scientific and technological studies. The quality and reproducibility of formulated processes depends on them. The manning of these laboratorics by pharmacists and engineers is supported by a highly sophisticated analytical research and development group.

While the pharmacists are spending their time developing biologically optimum qualitative and quantitative formulations of the drug that represent the best possible delivery systems; the analytical chemist is devoting his efforts to the development of specifications and test methods for all inert excipients, raw materials, bulk drug and finished pharmaceuticals, including statistical sampling procedures and treatment to give significance and reliability to test results.

To carry out this latter assignment, highly sophisticated analytical equipment in specially designed laboratories is made available to analytical development chemists. Their task is:

1. to define the appropriate quality of raw materials used in the manufacture of the drug substance and the dosage form and to develop test procedures to insure that such criteria are routinely met;

2. concomitantly, to provide technical expertise to the production and quality control operations in the resolution of problems that may arise in the manufacture and testing of drug products;

3. to work out in cooperation with the chemical and pharmaceutical development staff in-process controls that are so necessary for establishing the uniformity of production from batch to batch and the absolute integrity of the product; and

4. to develop chemical, physical, biochemical and biological methods that accurately measure the stability of both the drug substance and the pharmaceutical formulation. (Mislabeling will arise if a product goes out of specification during its specified shelf life.)

Finally, the pharmacy R&D group, with the analytical R&D staff, and with input from the production and marketing departments, finalizes the product development picture by:

1. establishing the optimum appearance of the finished product—this includes size, shape, color, taste, hardness, viscosity, injectability or other quality characteristic pertaining to product integrity and consumer acceptance;

2. establishing methods, facilities and controls to be used during manufacture, processing and packaging to assure maintenance of identity, strength, purity and homogeneity;

3. reconciliation of all data relative to the manufacture of the product. This completes the technical input.

These objectives are the responsibility of those development groups that are concerned with the technical development of the product. If done properly, the production and guarantee of the integrity of our products is enhanced immeasurably. This, however, is not the end of the technological story.

Package Development

Last, but certainly not least, is the area of package development. Here the objectives are to provide:

1. optimized package size and functional design;
2. specifications and test methods to assure packaging material compatibility with drug formulation;
3. label and literature design for the particular package size;
4. specifications for cartons and shippers to assure proper protection for the primary package; and
5. optimum stability between product and primary package.

This area concerns package development in cooperation with other development and control groups, and in our scheme of things cannot be minimized even though the product, not the package, is of primary interest.

The problems that can arise through package and product interaction, mislabeling, etc. are legion, and all come under the heading of CGMP with the same legal and medical difficulties that arise from product deterioration or increased toxicity.

Although package design is treated with considerable care during the initial period of testing and production, attention to this area may be minimized, if not eliminated, as a potential source of difficulty as time passes. This should not occur, since the quality of the primary container, especially with plastics, can change with time just as the drug itself.

Assuming all work in clinical, medical, product, and package development has been done and a product has been judged to be safe, unadulterated and efficacious within the framework of the new drug regulations, government approval of an NDA can be expected.

Approval, however, is not a guarantee of the quality of the product. To repeat a much used phrase, government cannot legislate nor regulate the quality of design and assurance into the product any more than quality control can put quality into a product by testing.

The absolute quality of the design is determined by the competence of the work during the design phase by all scientists and technologists involved in research and development.

This assurance is dependent primarily on the degree of professional commitment to the task of designing safe and effective drugs. Secondarily, but equally important, is day-to-day vigilance by all company personnel in seeing that all R&D and production programs are carried out with meticulous care.

The highest possible level of CGMP and quality control *must be self imposed* by pharmaceutical manufacturers, and the only assurance granted to the ultimate user is the competence and care of all the people involved—R&D, quality control, production, and regulatory groups—government regulations notwithstanding.

To summarize, *quality assurance* is the guarantee that the efforts of every professional have been extended to produce a quality product.

To insure this, what are some of the things that the industry can do and has done?

1. A thorough on-the-job training of personnel.

2. A feedback of quality control data to manufacturing personnel and management by way of reject and complaint reports; and evaluations of frequency of failure of raw materials and finished products to meet prescribed specifications for first analysis.

3. Involvement of nonprofessionals in observations and the collection of data used to measure product and production quality.

4. Involvement of all R&D professionals in the continuous revision and updating of personnel requirements, facilities, procedures and systems.

5. Development of long-range efficiency quality control programs.

6. Use of statistical studies of quality control results, along

with feedback of information to appropriate operating groups as well as to management.

7. Routine and regular inspection of all operating areas covered by CGMP regulations by personnel not reporting directly to production management.

8. Use of a quality alert reporting system to production and quality control that anticipates difficulties before they develop.

9. Review and reconciliation of all experimental and production data before product release.

10. Continually updating those systems and procedures for recording and review of all production and quality control data.

11. A comprehensive and exhausting laboratory testing program. This is the final stage prior to release of a product for human use and is the last opportunity to insure, as much as humanly possible, the identity, strength, quality, purity and pharmaceutical elegance of all drugs.

While this list is not complete, it underscores the major changes that have taken place at all company levels and the effectiveness of the development effort on a company's ability to achieve realistically the goal of providing a product meeting all stated claims.

Suggested Reading

Federal Food, Drug & Cosmetic Act as amended January 19, 1975. Superintendent of Documents, United States Government Printing Office, Washington, D.C. 20402.

Fleming, R., *Scale-Up in Practice.* New York: 1958, Reinhold Publishing Corp.

Hahn, Peter A. *Chemicals from Fermentation.* New York: Doubleday, 1968.

Killeffer, David H. *Chemical Engineering.* New York: Doubleday, 1967.

Levenspiel, Octave. *Chemical Reaction Engineering.* New York: John Wiley and Sons, Inc., 1962.

Schweyer, Herbert E. *Process Engineering Economics.* New York: McGraw-Hill Book Co., Inc., 1955.

CHAPTER 3

Discovery To Therapy

D. I. Weisblat

Perhaps the poet James Russell Lowell somewhat overstated the theme of this chapter some years ago when he wrote, "if the works of the great poets teach anything, it is to hold mere invention somewhat cheap. It is not the finding of a thing, but the making of something out of it after it is found, that is of consequence." (From *My Study Windows: Chaucer* 1871 by J.R. Lowell.) Lowell was not a scientist. One can question his use of the word "cheap" in this context, but it is indeed "the making of something" out of an invention or observation that benefits mankind.

We credit Banting and Best for discovering insulin, not because they were the first to demonstrate a factor from the pancreas that lowered blood sugar, but because they were the first to develop an extract of this organ for practical use in treating diabetes. Although Fleming discovered the germ-killing ability of penicillium mold extracts, Chain and Florey shared the Nobel Prize with him because their developmental work made penicillin available for the crucial clinical tests that established its utility. Following this demonstration, there was a massive mobilization of the fermentation and pharmaceutical industries to produce the miracle drug in quantity. There is no question that this developmental effort went far beyond the capabilities of the academic institutions where the basic discovery was made.

Academic laboratories play a remarkable role in fundamental discoveries, while industrial laboratories have facilities, equipment and manpower to take a discovery in the field of medicine and develop it to the stage of therapeutic utility. The ability of the industrial laboratory to produce, in quantity, material present in animal tissues in exceedingly minute amounts is the major factor in the academic-industry cooperation presented as a case history in this chapter.

Case History of a Developmental Program

The prostaglandins are a group of closely related, hormone-like substances uniquely different from the steroid and polypeptide hormones as illustrated in Figure 1.

The prostaglandin or "PG" story began in the early 1930s when a gynecologist and a physiologist, Kurzrok and Lieb, found that human semen produced either relaxation or contraction in isolated strips of human uterus. Later, both Goldblat in England and von Euler in Sweden published studies on the striking biological activity of human seminal plasma and of extracts of the seminal vesicular gland of sheep. Both were able to show that the biological properties of these materials could not be accounted for by any of the compounds known at that time. Von Euler pursued the characterization of the novel material and in 1935 coined the name "prostaglandin (PG)." By a fascinating quirk of fate, this work of von Euler's appeared on the same page of the *Klinische Wochenschrift* that carried the report of Butenandt, which named the important corpus luteum hormone "progesterone."

For almost twenty years the PGs were all but forgotten; very little research effort was expended. Von Euler, however, retained his interest in this field, and in the early '50s he encouraged a brilliant young colleague, Sune Bergstrom, then at the University of Lund, to resume efforts to purify and determine the structure of the PGs.

The Upjohn Company became involved with the PGs in the fall of 1957 when the author, by chance, wrote to Professor Bergstrom to ask some questions regarding a natural product program under way in the Upjohn laboratories. After replying to the questions in the letter, Bergstrom remarked that he was then much excited about, "prostaglandin of sheep prostate glands that we have separated into at least two factors, one of which is crystalline—a few mg.—with a specific activity approaching that of acetyl choline—but it is nitrogen-free and we

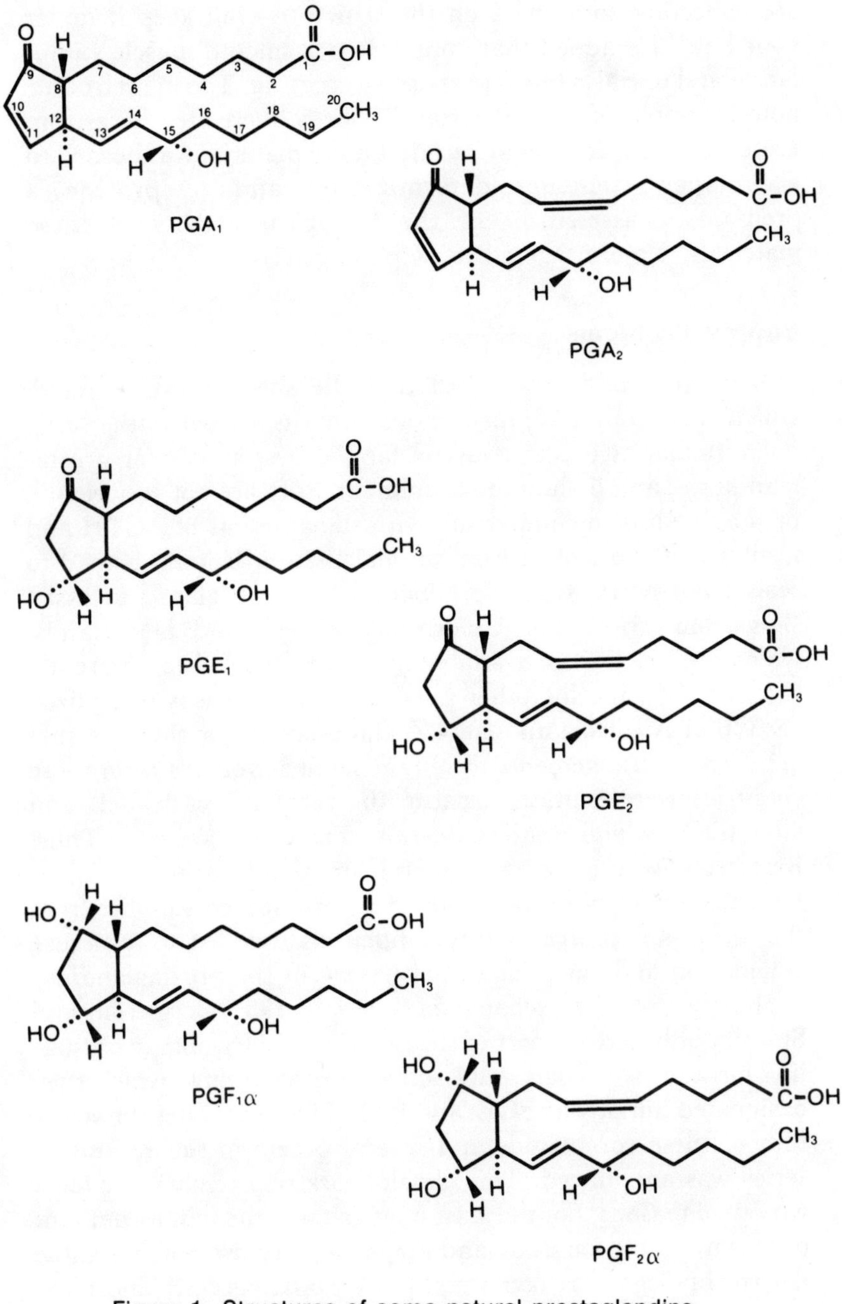

Figure 1. Structures of some natural prostaglandins.

are collecting some data on the structure—but keep it under your hat." He added that "one factor is smooth muscle stimulating and one also blood pressure decreasing. The main trouble now is supply of raw material." Shortly thereafter Bergstrom visited Upjohn. It was obvious that more material was needed to accomplish structure determination and to provide a preliminary assessment of the biological activity of these materials. Upjohn agreed to help Bergstrom.

Supply Problems

Bergstrom had established that the sheep prostate gland, which was really the prostate plus what is known as the seminal vesicular structure. was by far the best source for prostaglandins. Table 1 shows that animal tissues are not very rich in prostaglandins in comparison with the amounts of ACTH and insulin one can isolate from animal tissues. Upjohn helped to search out every available kilogram of sheep glands to assist Bergstrom's program. Unfortunately, beef and hog glands, which are much more available than those from sheep, were not suitable. Considering other possible sources, it was recognized early that Australia and New Zealand are major sheep-raising areas. But arrangements to obtain seminal vesicals from these countries were fruitless because the rams are castrated soon after birth, which inhibits development of the vesicals. Thus, Bergstrom was forced to depend for glands primarily upon Iceland, the Scandinavian countries, and limited supplies from the U.S. as a source of the compounds that led to structure elucidation and early biological testing of the prostaglandins.

Shortly after the exchange of letters in 1957, Bergstrom and Sjovall published a report of the isolation of a few mg. in crystalline form of two biologically active prostaglandins, which they designated initially as PGE_1 and PGF. Five years later the structure of these compounds and several others in the E_2 and E_3 series was announced. This classic work was remarkable for a variety of reasons, not the least of which was the minute amount of PGs available for study and the exemplary use of remarkable new technology that combined gas chromatography and mass spectroscopy for structure analysis as well as, finally, X-ray crystallography to determine the spatial arrangement of these structures. The family of PGs continues to grow as new members are isolated and characterized. They represent a family of compounds closely related as to chemical structure (see Fig. 1) but with quite distinct biological effects.

It soon became obvious that the continued collection of sheep glands would never yield sufficient material for definitive biological testing. It also was apparent that the potential of the PGs in therapy at this stage was certainly a very big question. About all that had been established was that these substances had a fascinatingly unique structure with an unbelievable mg. potency in causing constriction of smooth muscle and exerting vasodepressor activity. The potential for synthesis was carefully considered once the structures were known, and it was evident that the problem was going to be difficult because of the complexity of the molecules. Chemical instability of the PGEs and the PGAs also raised a dilemma.

TABLE 1

Yields of hormones from natural sources

Hormone	*Source*	*Human Doses/Kg.*
ACTH	Hog Pituitaries	4,000-5,000
Insulin	Beef Pancreas	100 - 200
PGF_2-alpha	Sheep Vesicular Glands	1/5 to 5

Biosynthesis

Fortunately, one could refer to a chapter from Upjohn's past experience in the steroid field—biosynthesis. It happened that our chemists and biochemists were asking themselves the questions: How might the PGs be formed in man, and what were their biochemical precursors? A trio of imaginative scientists, Phil Beal, Gun Fonken and John Pike at Upjohn, in discussing potential precursors for the PGs, visualized the possibility that a natural fatty acid found in the body, arachidonic acid, might under certain chemical or enzymatic conditions be converted to a PG. Other scientists were independently thinking very much along the same lines, and the first published reports of such conversions appeared from Bergstrom and from David Van Dorp in Holland.

These bioconversions are not restricted to the use of vesicular glands, as the necessary enzyme is widely distributed in other body tissues. Extensive development of this conversion was undertaken and a pilot plant was set up to produce gram quantities of PGs. This biosynthetic approach was a godsend which kept the research alive because it provided materials for nearly all the PG studies on a worldwide basis for almost five years.

The biological studies carried out in this period in many different countries with PGs furnished largely by Upjohn, and the ensuing early clinical studies in England and Sweden, supplied enough solid information to convince us of potential therapeutic utility. We were quite excited by the action of the prostaglandins on the cardiovascular system—with obvious implications to hypertension and effects on the female reproductive system, as well as fertility control. However, many other scientists and industrial laboratories were discouraged by the multiplicity of pharmacological actions and apparent lack of specificity demonstrated by the PGs.

Bioconversion, however, was not a cheap process. Supplies of PGs were still scarce and terribly expensive, with costs running in excess of $2,000 per gram or $900,000 per pound. Several process developments of major importance occurred about this time to break the supply bottleneck. Total synthesis methodology developed in our laboratories and elsewhere was an important factor together with a quite unexpected new development.

Prostaglandins from the Sea

This fascinating part of the PG story began in April, 1968, when we were informed that Professor A.G. Weinheimer at the University of Oklahoma had discovered a new natural prostaglandin source. A brief survey of Weinheimer's publications indicated that his research program dealt primarily with natural

Figure 2. A soft coral source of prostaglandins, *Plexaura homomalla.*

products from marine organisms such as coral. We supplied him with prostaglandin reference samples. The remarkable fact, soon established, was that his material, although chemically very much like PGA_2, was without the dramatic blood pressure lowering effect of PGA_2. In December, 1968, we received a sample of Weinheimer's PGA_2-like material and soon learned through physical-chemical studies, as had Weinheimer, that this was the "15-R" structural variant of the natural material. We also learned that the biological source for this prostaglandin was a Gorgonian soft coral called *Plexaura homomalla* or, commonly, sea whip, illustrated in Figure 2. This coral, collected off the Florida coast, contained prostaglandins in the cortex of the coral in very remarkable concentration compared with quantities found in mammalian tissues. Prostaglandins or their esters were found to be present in the air-dried coral at levels of from 1-3%. What physiological role these compounds play in this particular coral is completely unknown at this time.

Plexaura homomalla is distributed in considerable quantity in the Florida area, and arrangements were made to collect it with full awareness of potential ecological problems. Actually, a moderately heavy storm breaks loose much more coral than that collected throughout the program. Chemists worked out improved extraction techniques for the prostaglandins present in the coral and also devised the chemical steps essential to convert the unnatural "15-R"-isomers to natural (or 15-"S") physiologically active prostaglandins such as PGE_2 and PGF_2-alpha.

A survey of *Plexaura homomalla* growing in other areas of the Caribbean established the interesting fact that some specimens contained PGs of the natural 15-"S" configuration. In many of these, no 15-"R" prostaglandins can be detected by any analytical technique. Some individual colonies are found that contain mixtures of both the natural and unnatural compounds. Coral served as an interim source for PGs for research, but total synthesis now meets commercial needs.

Biological Activities—Clinical Applications

As available supplies of the prostaglandins have increased, the range of biochemical and biological tests and systems in which they have been studied has also grown. Two points are important: First, prostaglandins are among the most potent physiologically active materials known since as little as five to ten nanograms (a nanogram is one-billionth of a gram) of PGA_2

TABLE 2

Blood Pressure Effects of Prostaglandins

Prostaglandin	*Man*	*Monkey*	*Rat*	*Cat*	*Rabbit*	*Dog*
PGA_1	–	–	–	–	–	–
PGA_2	–	–	–	–	–	–
PGE_1	–	–	–	–	–	–
PGE_2	–	–	–	–	–	+
PGF_1-alpha	+	+	+	–	–	+
PGF_2-alpha	+	+	+	–	–	+

will produce a fall in arterial blood pressure in the cat; second, the individual prostaglandins have widely different biological effects and also show marked species differences. As shown in Table 2, prostaglandins of the E and A series depress blood pressure in a variety of species while those of the F series, with the exception of the cat and rabbit, show pressor responses.

These compounds rapidly stimulated research in a great number of completely unrelated fields, and requests for materials for laboratory studies poured in from scientists investigating endocrine physiology, lipid metabolism, ocular physiology, lung dynamics, central nervous system function and reproductive physiology, among many others.

Particularly striking and exciting are the effects of the prostaglandins in the area of reproductive physiology. Intravenous administration of very low doses of PGE_2 and PGF_2-alpha leads to stimulation of the uterus. This has led to their use, first by Professor S. Karim in Uganda, for labor induction in several thousand women. Oral administration of PGE_2 and PGF_2-alpha has also been reported, first by Professor Karim, as an effective means for inducing labor. The additional clinical studies needed to assess the value of prostaglandins relative to presently accepted methods, and to devise correct dosage schedules and best routes of drug delivery, are in progress in several countries, including the U.S.

Also, studies at Upjohn by Pharriss and by Kirton showed that PGF_2-alpha was effective in preventing and terminating pregnancy in a variety of animal species, including the monkey. As a consequence studies in women were undertaken. Two groups, Bygdeman and Wyqvist in Stockholm and Karim and Filshie in London, simultaneously in 1970 reported in *Lancet* successful abortions with PGF_2-alpha.

The various routes of administration of the prostaglandins for either labor induction or therapeutic abortion have received much attention, especially since these agents produce such a variety of pharmacological effects. Since the goal for these indications is to stimulate contractions of the uterus with minimal action on other systems, the approved route of administration for PGF_2-alpha abortions in the U.S. is intra-amniotic or directly into the amniotic sac. Vaginal suppositories of PGE_2 effectively produce abortions with relatively few side effects. Those occasionally observed are nausea, vomiting and diarrhea produced because of the stimulant effect of prostaglandins on the gastrointestinal tract.

As yet, the role of the PGs in the normal reproductive process is not fully understood, nor is there a clear picture of the hormonal interrelations involved in prostaglandin-induced abortions. It is apparent, however, that there is worldwide interest in the substances because of the overpopulation problem.

Possible implications of prostaglandins in hypertension became evident when J.L. Lee's group in St. Louis established that a blood pressure lowering lipid called "medullin" from the kidney medulla was indeed PGA_2. As in the coral story, comparison standards supplied by Upjohn played a major part in establishing the structure of this material.

Since the PGAs stimulate salt and water excretion in addition to lowering blood pressure, they are attractive candidates for clinical evaluation, and exploratory studies are in progress. This is a most promising area for continuing research.

Andre Robert in the Upjohn laboratories has shown that prostaglandins inhibit gastric acid secretion, and more likely, acid formation. It was possible to demonstrate in dogs that PGE_1 or PGE_2, given intravenously, inhibited gastric acid secretion stimulated either by histamine or food intake. This work has been extended to a study of the effect of PGs in experimental ulcer formation in the rat. For example, gastic ulcers were produced by the administration of corticosteroids, and formation of this type of ulcer could be inhibited by PGE_1. Evaluation of the potential for these agents in the treatment of ulcers in man is underway.

Another area of potential application of the PGs is in the treatment of bronchial asthma. Prostaglandin aerosols, particularly A_1 and E_1, have shown powerful bronchodilator activity in preliminary human trials. On a weight basis, the PGs were more potent than currently available synthetic agents.

There are many other aspects of the physiological actions of the PGs, which are beyond the scope of this presentation. Table 3 summarizes the most important potential applications of the PGs and lists a number of other possibilities of a more exploratory nature.

TABLE 3

Potential clinical applications of the prostaglandins. In several countries prostaglandins are now on the market for labor induction and/or abortion.

Labor Induction	Menstrual Regulation
Abortion	Male Infertility
Asthma	Blood Platelet Preservation
Hypertension	Ulcers (Stomach and Duodenal)
Shock	Peripheral Vascular Diseases
Nasal Decongestant	Dermatology

Thus far only the natural prostaglandins and their physiological properties have been discussed. Based on experience in the steroid and other fields, it was decided very early that analogs might be of value. To date some 500 PG entities have been made. Quite a number of these analogs might have been studied to determine the effect of structural modification on the nature and duration of biological activity. Figure 3 illustrates inactivation of a prostaglandin by the enzyme system C-15-Dehydrogenase. Analogs have been made with a methyl group at Carbon-15 which blocks this reaction. One would hope, therefore, for retention of activity and prolonged action; and indeed, 15-methyl-PGF_2-alpha and 15-methyl-PGE_2 are from five to ten times as potent in the monkey as the parent compounds and show prolonged activity in a variety of test systems.

Sample Distribution

A feature of the prostaglandin research program at Upjohn has been the distribution of samples of these compounds without charge to interested investigators in many fields. Some twelve thousand samples have been distributed to some six thousand investigators throughout the world. This approach

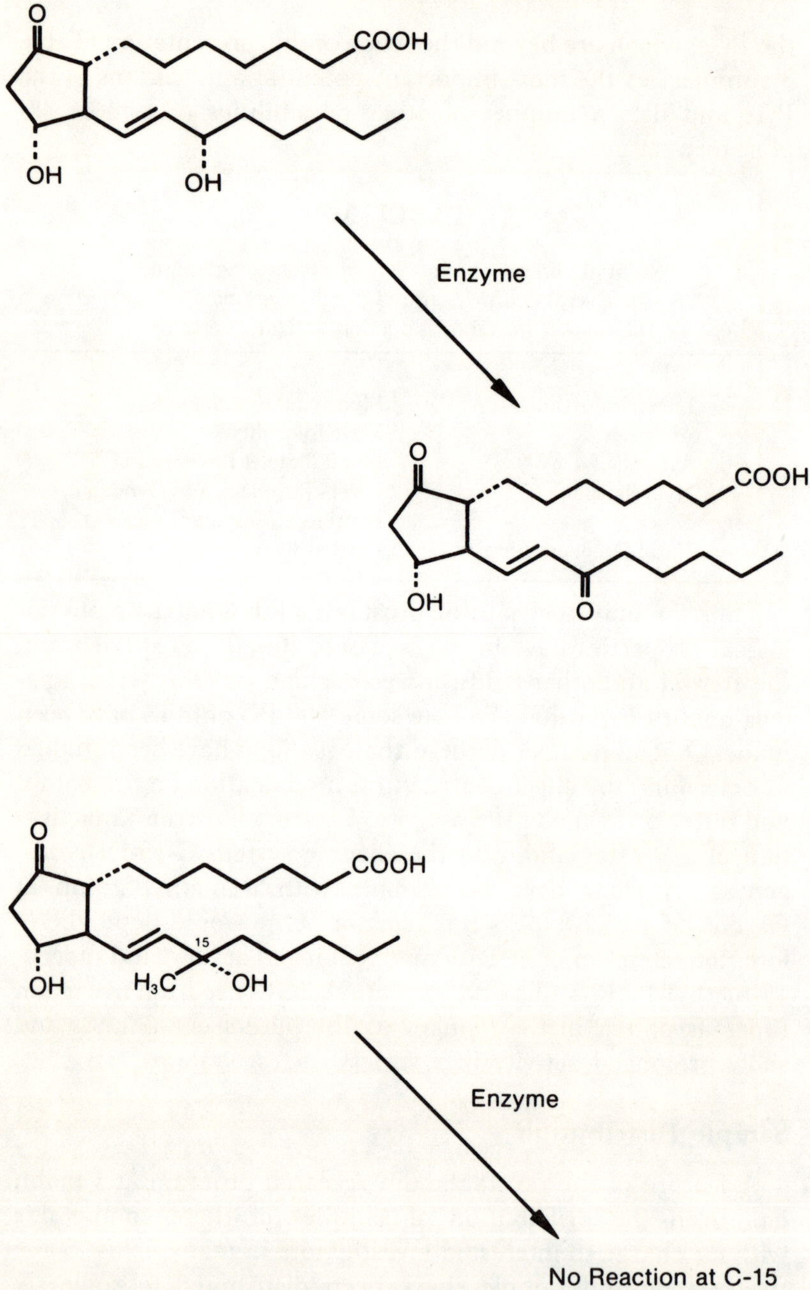

Figure 3. Blocking of enzymatic inactivation of prostaglandins.

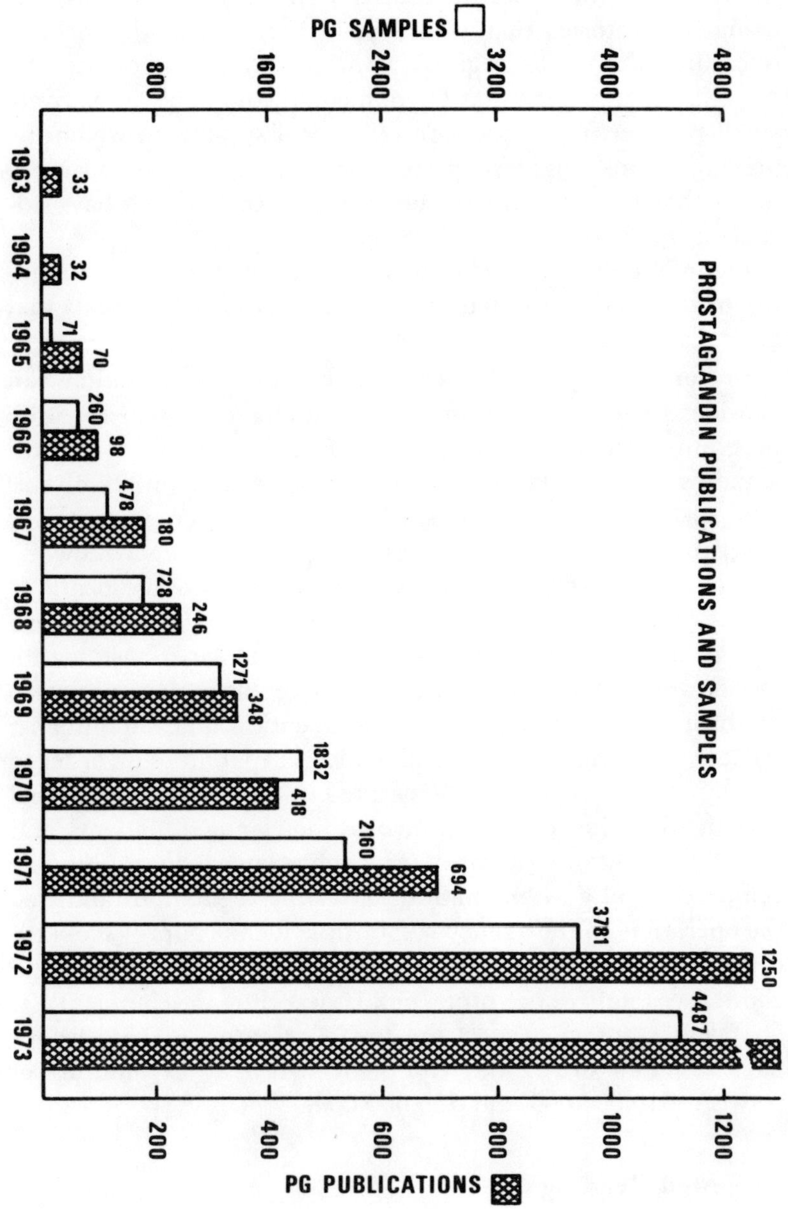

Figure 4. Prostaglandin publications and sample distribution.

was taken in the belief that seeds were being planted in academic laboratories that would yield a return in the form of scientific information regarding the potential of these agents. Also, it was apparent that Upjohn could study only a relatively small number of the many physiological systems in which the prostaglandins might play some role. It is the conviction of many that these seeds have bone fruit manyfold and have advanced the field by five to ten years.

Figure 4 indicates a very good correlations between the number of samples distributed and the number of publications that have discussed these substances.

In summary, a pharmaceutical research organization can provide the following unique contributions to a joint research program with academia: (a) the drug, in this case, prostaglandins; (b) standardized toxicity studies in a variety of animal species to guarantee safety and meet requirements of government regulatory agencies; (c) pharmacy know-how essential to the development of a stable, accurate dosage form; (d) access to human volunteers for tolerance studies to determine a dosage range prior to study in patients; (e) experience with a broad range of medical people knowledgeable in drug testing to establish a protocol for the first therapeutic studies in humans; (f) data handling and statistical analysis capabilities, in order to evaluate the extensive data generated in the course of a clinical investigation that can often involve thousands of patients and literally mountains of data; (g) people knowledgeable in the complexities of governmental drug testing regulations and also the operation of the agency responsible for the approval of new drugs; (h) concentrated and directed multidisciplinary research efforts by a staff whose profession is new drug development; (i) financial resources to meet the cost of research as the development of a new drug today can often involve an expenditure of between $10,000,000 and $20,000,000.

Suggested Reading

Bylinsky, G. Upjohn Puts the Cell's Own Messengers to Work. *Fortune*, June, 1972.

Cuthbert, M.F. *The Prostaglandins*. London: William Heinemann Medical Books, 1973.

Higgins, C.B., and Braunwald, E. The Prostaglandins, Biochemical, Physiologic and Clinical Considerations. *The American Journal of Medicine*, 1972, **53**, 92-112.

Horton, E.W. *Prostaglandins*. New York: Springer-Verlag, 1972.

Karim, S.M. *Prostaglandins: Progress in Research*. New York: Halsted Press, Division of John Wiley & Sons, Inc., 1972.

Lee, J.B. *Perspectives on the Prostaglandins*. New York: Medcom Press, 1973.

CHAPTER 4

Splitting the Antipodes

Ira Ringler

Pallor and an alabaster skin were important aspects of womanly beauty in the nineteenth century. But the whiteness was all too often a signal of impending death as well. Surrounded by her friends in a Paris garret, Mimi, the pale young heroine of Puccini's *La Boheme*, died very quietly as her lover, Rudolfo, wept at her side. Death for the languishing beauty of this Italian opera was not violent. It was a gradual slipping away . . . a feeble surrender to the fashionably destructive disease known as consumption.

By 1905 consumption was the leading cause of death in the United States, particularly in children and young adults. Today we call it tuberculosis, or TB, and most people think, because of modern medicine, that the problem of TB has been solved. This is far from true. The World Health Organization estimates that 20,000,000 people in the world today have TB, and 2,000,000 die of it each year. Despite remarkable medical and social advances in recent years, the battle is far from over. In 1972, 4,000 Americans died of the disease and 33,500 new cases were reported. Approximately seven percent of the population of the United States (15,000,000 people) harbor tubercle bacilli.

Tuberculosis is a constant threat; some experts predict that it will remain so for at least the next fifty years. Crowding, poor sanitation, lack of proper hygiene and poor nutrition . . . all contribute to a high incidence. Society creates the environment for the infection to exist . . . as it waits for the opportunity to break out, incapacitate and kill people who barely recognize its presence.

The term "tuberculosis" is derived from the shape of the lesion formed by the invading bacillus. The organism that causes

TB is called the "tubercle bacillus." This is a one-celled microbe shaped like a rod or tube, which may occur in dust and be blown about in the air. When inhaled or taken with food, the bacilli may live and multiply in the soft, moist tissue of the lungs. They can also infect almost all other tissues and organs of the body. Body cells become sensitive to protein products of the bacilli, causing irritation and even cell death. Minute abscesses develop. White blood cells cluster around the invaders and form a strong coating. In this manner the body encapsulates the bacilli to prevent their spread. The TB bacillus is one of the toughest disease-causing organisms in existence.

For a time, medical researchers hoped to find a treatment that would wipe out TB within 10 to 20 years. Vaccines, although quickly successful in virtually eliminating diseases such as diphtheria, have not provided the answer to TB. No prospects for eradicating it are held in high hopes now. The present goal is control through the constant development of new, safe and more effective anti-TB drugs.

Drugs currently used to treat TB include streptomycin, para-aminosalicylic acid (PAS), thiacetazone, viomycin, isoniazid, pyrazinamide, cycloserine, kanamycin, ethionamide, thiocarlide, capreomycin, ethambutol and rifampicin. The chemical structures of twelve anti-TB drugs are given in Figure 1. Among these, the three most efficacious—isoniazid, ethambutol and rifampicin—were developed by the drug industry. (Capreomycin is a polypeptide antibiotic and is not included in Figure 1 since it is a complex of four microbiologically active components that have been characterized only in part.)

Myambutol®, ethambutol, one of the most effective and safe anti-TB drugs, was developed by American Cyanamid Company as the result of an extensive search that spanned 24 years. Its development illustrates the complexity of a scientific effort that began in 1944, demanding the many skills of chemists, biologists, pharmacologists, pathologists, clinicians, mathematicians and other highly trained scientists. The project required an interdisciplinary team organized to reach a well-defined therapeutic objective—the kind of capability that has evolved over a number of years in the research-oriented pharmaceutical industry. The use of animal models of human disease to detect the therapeutic activity of chemical compounds is one of the most important factors responsible for the advances in medical therapeutics during the past 30 years. Scientists in the industry have devoted much effort to improving the efficience and reliability of animal screens used to find useful new drugs.

In 1944, only streptomycin was available for the treatment of TB, and it had its drawbacks. That year, the Lederle Laboratories Division of American Cyanamid Company established an animal screening program to find new anti-TB drugs, a test that now seems slow and cumbersome. This early effort was suitable for screening about 500 new chemical compounds per year. Each new compound was tested for anti-TB activity in 10 guinea pigs, and because of the relatively large size of this laboratory animal, about 10 grams of the substance to be tested were required. After each test, the guinea pigs were autopsied and eight organs of each animal were weighed, examined and graded for pathological involvement. It was a long and tedious process. The extent of the anti-TB activity was measured by comparing the tissues of treated guinea pigs with untreated control tissues.

TABLE 1

The Development of
Lederle's Antituberculosis Screens

	Requirements for Each Compound			
Year	*Grams*	*Animals*	*Observations*	*Test Capacity (Compounds/Year)*
1944	10	10 Guinea Pigs	Examine 8 organs	500
1945	2	10 Mice	Examine lungs	1,000
1956	1	5 Mice	Survival	2,000
1961	0.1	2 Mice	Survival	4,000
1963	0.05 (5 Compounds)	2 Mice	Survival	20,000

The original Lederle test was vastly improved by continual

Figure 1. Chemical Structures of Twelve Antituberculous Drugs.
(Excepting Rifampicin, all structures were redrawn with permission from The Merck Index, Seventh Edition, copyright 1960 or Eighth Edition, copyright 1968, Merck and Co., Inc., Rahway, New Jersey, U.S.A. The structure for Rifampicin was redrawn from Abstract 81:58307u, Chemical Abstracts, copyright 1974, American Chemical Society, Chemical Abstracts Service, Columbus, Ohio, U.S.A.)

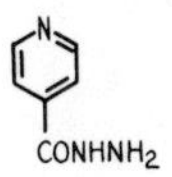

Isoniazid

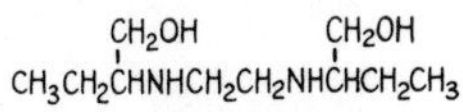

Ethambutol

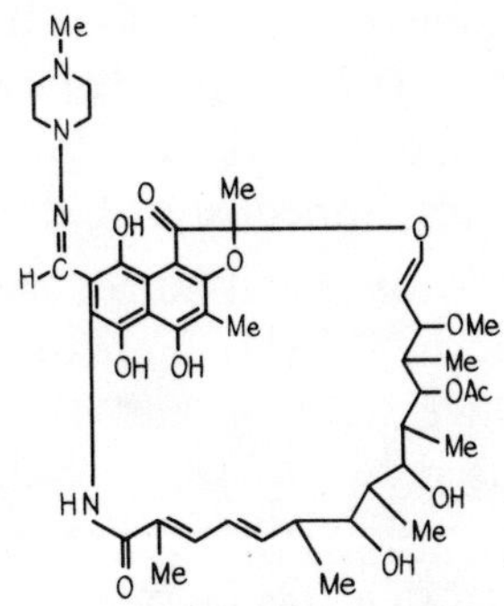

Rifampicin

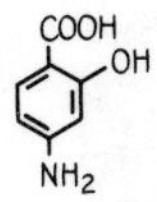

p-Aminosalicylic Acid

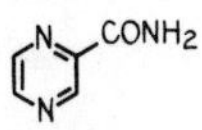

Pyrazinamide

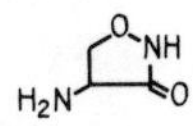

Cycloserine

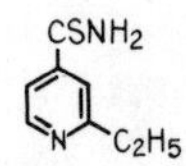

Ethionamide

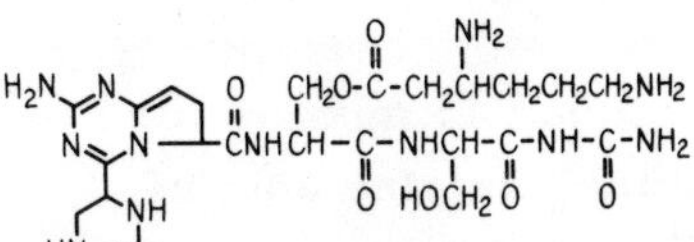

Viomycin

Thiocarlide

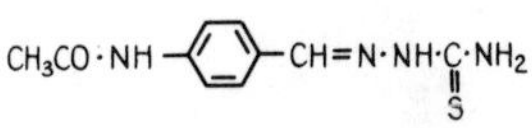

Thiacetazone

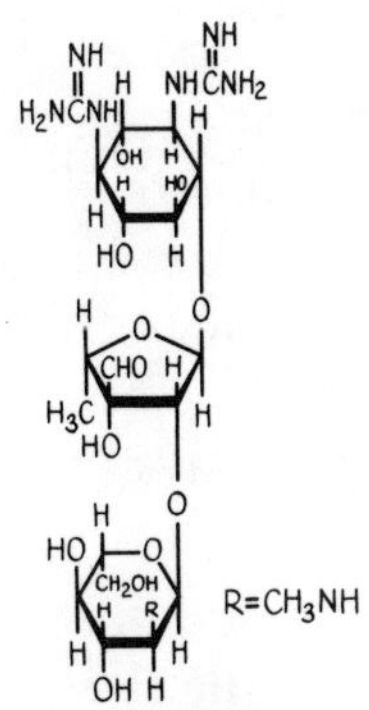

Streptomycin

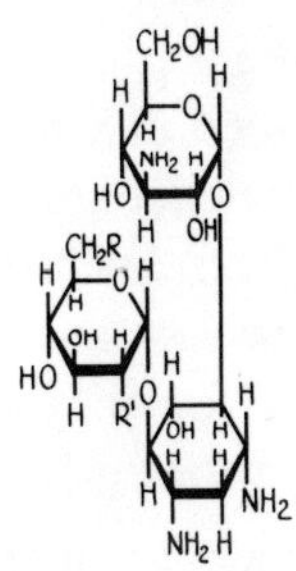

	R	R'
Kanamycin A	NH_2	OH
Kanamycin B	NH_2	NH_2
Kanamycin C	OH	NH_2

experimentation and change (Table 1). Mice were substituted as the experimental hosts for the TB bacillus, and eventually it was possible to screen 20,000 compounds each year. The test finally adopted utilized only one twentieth of a gram of each new compound, and, even better, was set up to screen five compounds simultaneously. A description of it follows:

> A standardized inbred strain of female albino mice is used in all tests. Female mice are used exclusively to reduce fighting and the mortal injuries sometimes caused by fighting. A standard amount of a saline suspension of a human strain of the TB bacillus cultured in a special medium is introduced into the mice intravenously. Two infected mice are allowed for each group of five candidate compounds to be tested. Each group of five compounds is mixed with powdered mouse food. The diet mixtures are made up to contain 0.05 percent of each compound under test. If 100 infected mice are included in a single two-mouse screening test, 90 of these may be used for candidate compound mixtures, allowing tests of 45 mixtures or 225 compounds in all; the remaining infected mice include several untreated animals and animals treated with a standard reference compound (for example, isoniazid) for control purposes. Each group of two mice is placed in a separate cage and they are allowed to feed at will from containers filled with their special drug-diet mixture for a period of 14 days, beginning on the day they are inoculated with the TB bacillus. Survivors are allowed to feed on nontreated powdered food for an additional 14 days (a total of 28 days after inoculation).
>
> This type of test is designed to reveal the anti-TB activity of one or more of the five components in each diet mixture but it may also reveal the toxicity of one or more of the components at the 0.05 percent concentration. The infected nontreated control mice usually die within 14 to 21 days after the TB bacillus

> has been introduced into their bodies. If one or both mice in a treated group die before the first death in the infected controls, the mixture of compounds is rated toxic, and retests of each component must be made in subsequent tests at one or more concentrations in the diet to determine which component or components must be screened at concentrations lower than 0.05 percent. If neither mouse in a treated group dies before the first death in the infected, nontreated controls, but both die within the period that infected controls die, all of the components in that particular mixture are rated inactive against TB. If one or both mice in a treated group survive for a significant period of time after all of the infected, nontreated controls are dead, the mixture of compounds is rated active against TB, and each component is retested. If one or more of the individual components of an "active" mixture reveals activity in a repeat two-mouse test, additional testing is conducted, utilizing five- or ten-mouse groups and various doses and routes of administration. Accepted compounds are then further evaluated for efficacy and toxicity. Such compounds also serve as leads for chemical modification in the hope of developing drugs with improved therapeutic properties.

Over the years this screening test has been found to be highly reliable. It is capable of detecting a compound only one tenth as potent as the reference compound, isoniazid, with a statistical probability of 95 percent.

The efficacy testing phase of Lederle's TB research is performed by scientists and technical personnel working in masks and gowns in an isolated building with controlled air flow and other elaborate protective measures.

A large reservoir of new chemicals is needed for an effective screening program as well as a good animal test. The American Cyanamid laboratory file contains more than 100,000 chemical compounds, and new compounds are added at the rate of 5,000 annually. The compounds tested each year for anti-TB activity at the Pearl River, N.Y. laboratories are developed at different Cyanamid divisions as well as at Lederle. The file is computerized, and the chemicals are characterized so that specified structures may be selected quickly for testing should the random screening reveal a new structure of interest. Empirical or trial-and-error methods of searching for new leads among large numbers of chemical compounds contrast with selective screening or "rational" drug design. The latter is based on prior knowledge of mechanism of action—tailoring the synthesis of compounds along lines of certain basic concepts concerned with cell nutrition, enzyme action or pharmacological antagonism, and testing only these tailored compounds. Both approaches have resulted in useful chemotherapeutic agents, and, in combination, were employed by Lederle chemists, as will be demonstrated later.

During the operation of the Lederle anti-TB screens, a few compounds, found between 1944 and June, 1957, satisfied at least the minimal conditions of acceptance for further study. Among these was pyrazinamide, one of the first novel synthetic anti-TB drugs (Figure 1). It was ultimately found to be effective in human tuberculosis, but because of potential liver toxicity it was not used widely. However, some clinicians still regard pyrazinamide as a useful anti-TB drug. None of the other leads picked up prior to July, 1957, reached clinical trial, either because they were too toxic or were less effective in animal tests than other anti-TB drugs. Then, in July, 1957, after many thousands of compounds had been screened, potent anti-TB activity was discovered in a chemical called diisopropylethylenediamine (Fig. 2). The new lead compound was immediately striking because it was relatively simple, with a structure fundamentally different from that of other anti-TB drugs. It had originated at Cyanamid's Research Center in Stamford, Connecticut, where it had been made as an antioxidant additive for rubber. It failed as a good additive for rubber, but it was placed in the Company's master file and became available for all routine screening programs throughout the Company, including the anti-TB screen at Lederle.

$(CH_3)_2CH\text{-}NH\text{-}CH_2CH_2NH\text{-}CH(CH_3)_2$

Figure 2. Diisopropylethylenediamine

Enthusiasm was great at this point in the 13-year search, even among disappointment-hardened researchers. The anti-TB lead presented research management with a series of difficult decisions. Should this new substance be developed immediately? Should the chemists embark upon a program involving the synthesis of a series of chemical cousins or analogs? After all, there was always the possibility that an even more promising drug would emerge from such a program.

It was a tough decision. No compound quite so promising had been uncovered for many years. This new one would apparently be relatively simple to make, and Lederle could get an urgently needed new medicine to patients.

The question of safety was also a major consideration. Usually, anti-TB drugs must be administered to patients for one to three years, while most other antibacterial drugs usually need to be given only for several days or weeks. Would the new drug prove to be safe enough to be used in a patient continuously for several years? The answers—how effective and how safe—would not come quickly. A marketable drug was a long time off. One estimate was four to six years for the compound already in hand.

To move the project as rapidly as possible, the most comprehensive approach was selected. Work began simultaneously on development of the lead compound and on chemical studies to find possibly better analogs. It was retested again and again for efficacy in mice infected with the TB bacillus, using oral treatment, subcutaneous and intraperitoneal injections and various comparative tests involving reference anti-TB drugs such as streptomycin, p-aminosalicylic acid and isoniazid. Preclinical toxicity studies in various animals were conducted and, after two years of laboratory effort, the new lead compound was finally read for the start of clinical evaluation in TB patients; this was in May, 1959.

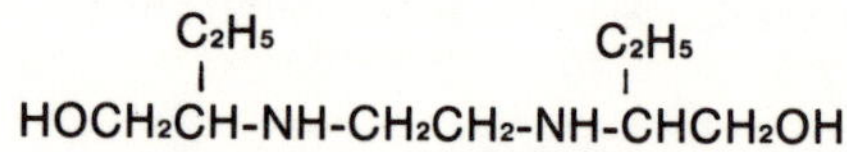

Figure 3. Hydroxybutylethylenediamine

But then, another element entered the picture. The research chemists, in their effort to produce agents with greater potency and lower toxicity than the lead compound, had synthesized 600 new structures. One of these, hydroxybutylethylenediamine (Fig. 3), was found to be about one-eighth as toxic to mice and about twice as potent in mouse tuberculosis as the original lead compound (Table 2).

TABLE 2

Effects of New and Original Ethylenediamine Leads Against Tuberculosis in Mice

	Dose, MG/KG/Day		
Compound	*Effective*	*Maximum Tolerated*	*Safety Margin**
Diisopropylethylenediamine (Original lead)	150	600	4
Hydroxybutylethylenediamine	88	4800	50

*Maximum tolerated dose/effective dose.

Again, it was a time for decision. Should one proceed with the original lead, or should the whole process of preclinical testing start again with the new compound?

It was decided to discontinue development of the original lead and proceed with the new compound. The results of 2 years of intensive animal safety studies, a large supply of the original compound and a wealth of information on processes and procedures for its commercial production were all set aside.

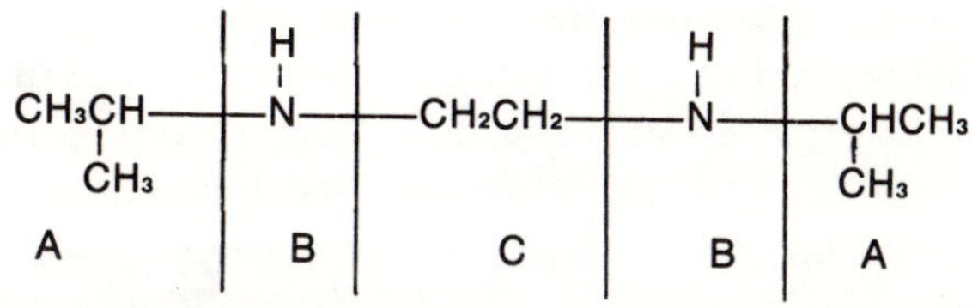

Figure 4. Schematic Diagram of the Structural Components of Diisopropylethylenediamine

Hydroxybutylethylenediamine had been made by research chemists adopting the strategy of molecular modification. The structure of the original lead compound could be divided into five components, as illustrated in Figure 4. Systematically, new compounds were made by substituting other chemical groups for one or more of those five components. For example, the A components of the original compound in Figure 4 are called saturated hydrocarbon or alkyl groups; it was found that relatively slight variations of the A components resulted in a loss of anti-TB activity. A small alkyl group branched at the carbon attached to component B was required for high anti-TB activity in mice. Variations of component B generally lowered or resulted in a loss of activity. Changing the length of component C by even one carbon atom brought about a loss of activity. Many other structural variations were tried, and it was found that the requirements for activity were remarkably strict. Then, an interest developed in the chelating potential of certain hydroxyalkyl ethylenediamines, and this led to the synthesis of hydroxybutylethylenediamine. A chelate is a compound having two or more points of attachment, with an atom of metal joined to form a ring-type structure. It is thought that chelation may produce chemotherapeutic activity by interfacing with one of a number of different enzyme systems, or it may have indirect effects such as facilitating cell penetration or absorption of a drug from the gut. Although a positive correlation between chelation and anti-TB activity is not certain, the possibility that such a correlation might exist influenced the strategy of preparing compounds, ultimately leading to hydroxybutylethylenediamine, the second highly potent anti-TB agent synthesized by Lederle chemists.

The next step in this complex series of events involved extensive efficacy evaluation of the new compound and studies of its pharmacological behavior in various animals to determine its safety . . . procedures that were all too familiar, since many of them had been carried out only a short time previously with the original lead compound. When the new compound was added to cultures of nine strains of the TB bacillus, seven species of bacteria besides the TB bacillus, and 11 species of fungi, only the TB bacilli were adversely affected. In animals, tests were conducted in experimental infections with three strains of the TB bacillus, eight other bacteria, three viruses, and five intestinal parasites. Efficacy was demonstrable only in the TB infections. The compound proved to be highly specific. Studies on its mode of action revealed that the drug enters the TB bacillus by diffusion and becomes bound to the cells; it inhibits the multiplication of the bacilli but only after a delay. Then the bacilli become deficient in ribonucleic acid.

Pharmacological and other preclinical studies in animals brought to light nothing that would contraindicate the use of the new compound in man. In November, 1960, it was given to man for the first time and was well tolerated. Then it was tried in critically ill tubercular patients who had not responded to other drugs. Good news! It was effective against TB in man. But there was bad news, too. There were reports of eye toxicity in some patients who were treated with high doses, although these symptoms disappeared when treatment was stopped. Still, this could have been the final blow to a project that had so far lasted

16 years. But hope was not lost. While the tests were going on in man, the chemists had pursued their work on hydroxybutylethylenediamine. They studied its characteristics more critically. The compound has two special carbon atoms. They are special because each is surrounded by the same series of four different chemical groups. There are two forms for compounds with this type of structure. One, called the meso form, is identical with its mirror image. The other form has two mirror image isomers, each of which has the same chemical properties but can be distinguished by the way a beam of plane polarized light is rotated as it passes through its solution. Our compound was of this latter type, known as a racemic mixture of two optically active components. Although optical isomers have the same chemical properties, they may differ in their biological activity. Louis Pasteur was the first to separate two crystalline forms of tartaric acid, identical in chemical properties but different in their light-twisting characteristics; he separated these forms with tweezers under a microscope.

Several methods were known to the Lederle chemists for attempting to separate the isomers, or antipodes of hydroxybutylethylenediamine. They could have employed fermentation techniques, since appropriate microorganisms might destroy one form and leave the other intact. Another method was to use enzymes that might break down one form but not the other. But still another method was finally chosen. It was known that acids combine with bases to form salts under some conditions, and salts of hydroxybutylethylenediamine prepared with an optically active acid were no longer physically identical and were separable by their differential solubility in methyl alcohol.

Quantities of the pure isomers were made for efficacy studies in experimental tuberculosis in mice. The results clearly demonstrated that virtually all of the anti-TB activity resided in the dextro form. This form had twice the safety margin of the racemate form (Table 3), indeed a gratifying finding. The decision then was made to throw all resources behind the new dextro form. The task of making large amounts of a pure dextro compound was in itself a formidable undertaking. Getting rid of the unwanted levo isomer in large-scale operations was a special

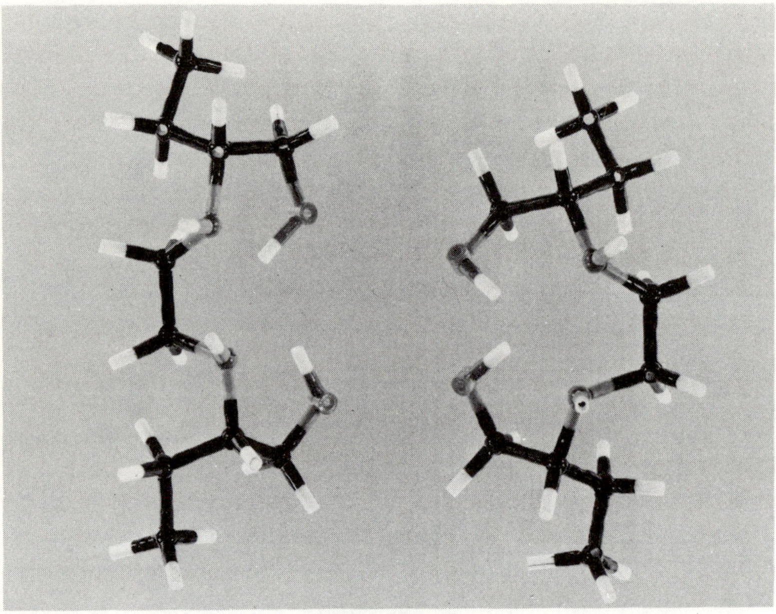

MESO FORM

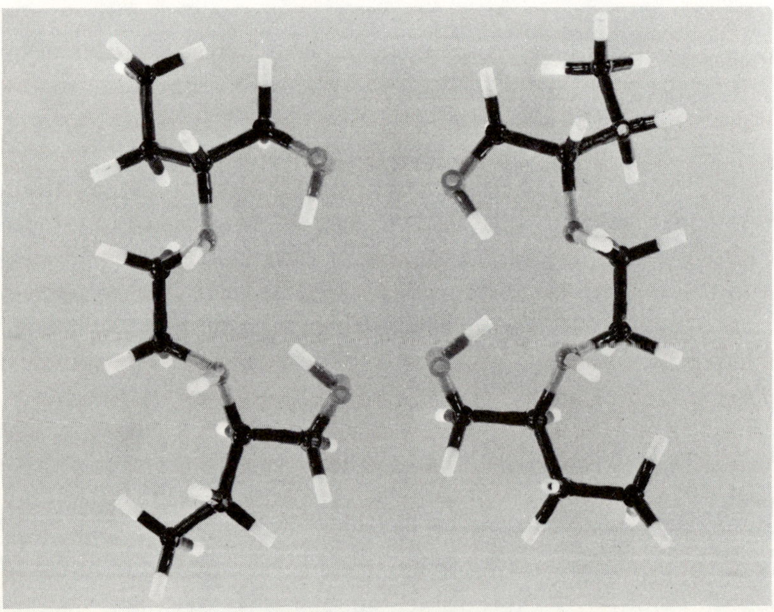

RACEMIC MIXTURE

Figure 5. Color Coded Skeletal Models illustrate isomers of hydroxybutylethylenediamine.

problem, and there was little in the scientific literature that might serve as a guide. Those who develop a process for making new drugs must be concerned with cost; otherwise, a proven chemotherapeutic agent might forever remain a laboratory curiosity. A high cost for any kind of therapy can severely limit the extent of its use. So, again the chemists were asked to start process development. After trial, error, and more trial, a method was developed for producing the dextro compound in commercial size batches and at a reasonable cost.

TABLE 3

Activity and Toxicity of the Antipodes of Hydroxybutylethylenediamine in Mice

Form of Compound	*Dose, MG/KG/Day*		
	Effective	*Maximum Tolerated*	*Margin**
d1-	88	4800	50
d- (Ethambutol)	44	4800	100
1-	Inactive	4800	—

*Maximum tolerated dose/effective dose.

Preclinical animal tests started with the new compound, which was given the name ethambutol (Myambutol® is Lederle's trademark). The extensive efficacy studies of ethambutol and its precursor, the d1 or racemic form, were carried out in mice by Lederle investigators, as were the preclinical toxicity studies that involved a variety of different laboratory animals. Then, another important link in the chain of events, which focused attention on the potential usefulness of ethambutol, was provided by studies undertaken in monkeys at the Christ Hospital Institue of Medical Research and the University of Cincinnati.

These investigations showed clearly that ethambutol had substantial antituberculosis activity against induced pulmonary tuberculosis in monkeys—one step nearer to human tuberculosis than the initial laboratory studies in mice. Although neurological symptoms, including visual disturbances, occurred in monkeys after prolonged high oral doses of ethambutol, the dosage required to produce severe symptoms was from 12 to 48 times as great as that which produced maximum benefit in established tuberculosis.

The first studies in humans with this new type of antituberculosis agent were conducted with the racemate form, but much more extensive investigational clinical studies were subsequently carried out with ethambutol. Preliminary clinical testing indicated that oral doses of the new drug were well-absorbed and well-tolerated. Cautious but definitive studies in TB patients showed that it was highly effective, especially in combination with isoniazid. Virtually no side effects were reported except, again, for a reversible abnormality in visual function. In those patients where visual disturbances were attributed to ethambutol, recovery generally occurred over a period of weeks or months after treatment was discontinued. Such disturbances were reported only from a small percentage of treated patients and were not considered sufficiently serious to shelve the drug, particularly since it was very effective against TB. What's more, in some well-controlled studies, the frequency and magnitude of decreases in visual acuity were no higher than among patients on regimens not containing ethambutol.

A list of all the physicians and their collaborators—bacteriologists, pharmacologists, chemists, biochemists, pathologists, laboratory technicians—involved in clinical studies of ethambutol, together with their contributions, would require many pages of print. It was necessary to determine safe doses for varying periods to find out whether side effects occurred, and their nature, during or following different dosage regimens. It was necessary to determine the effects of the drug alone and in combination with other anti-TB drugs, as well as effects in cases of tuberculosis resistant to treatment with other drugs. This required a multi-hospital, multi-investigator approach, involving an enormous number of people. For example, in one study in Japan, eleven different institutions or hospitals conducted clinical evaluations of ethambutol for more than two years; more than 200 patients were treated variously with ethambutol alone or in combination with other drugs, and nearly 100 additional patients were treated with other anti-TB drugs for control purposes. In this study, ethambutol was found to be effective in the treatment of pulmonary tuberculosis. Tubercle bacilli disappeared from the sputum soon after the start of treatment, and the incidence of side effects remained low.

In another illustrative investigational study, this time in the United States conducted with the collaboration of the Public Health Service, there were 27 cooperating clinical investigators from 13 states, working in 21 different hospitals or sanitoria. Their first published report listed 271 patients treated with ethambutol alone or in combination with other anti-TB drugs and 295 patients treated with other anti-TB drugs for control purposes. This work advanced the reputation of ethambutol as a promising drug for the treatment of tuberculosis.

There were many such investigational studies before the efficacy and safety of ethambutol was thoroughly established, and it won its niche among the important therapeutic weapons to fight one of man's most dreaded diseases.

By 1968, the long trek toward a successful anti-TB drug was nearing its end. Data had accumulated to the point where a quarter-ton truck was needed to carry to the Food and Drug Administration in Washington the many boxes of animal and human experimental test reports in support of a new drug application. The FDA approved the new drug for marketing in the United States in 1968. This was eleven years and three and a quarter million dollars after the discovery of the first lead compound in 1957, and twenty-four years after the beginning of the screening program in 1944.

Experience with Myambutol® ethambutol to date has fully confirmed the initial findings that it is one of the most effective and safe anti-tubercular drugs.

Patients have been successfully treated with Myambutol® in more than eighty-seven countries.

The successful outcome of the search for an advanced anti-TB drug depended on the interdisciplinary expertise of a well-organized team with the long-range commitment of time and funds.

Paul Ehrlich, considered the "Father of Modern Chemotherapy," recommended four indispensable prerequisites for the development of a new drug: *gehirns* (brains), *geduld* (persistence), *geld* (money) and *gluck* (luck). If Ehrlich were alive today, perhaps he would recommend a fifth—*glaube*, or faith.

Right now, many pharmaceutical companies are relentlessly searching for new forms of therapy to treat and prevent the many diseases that still maim, cripple and kill. In this context, it is fair to point out that for each success story of search and development there are many other sometimes exciting potential drug candidates that never reach the point of testing in human disease victims or, worse, that reach clinical testing but are rejected after extensive and expensive trials. Happily, Myambutol® did not fail.

Note: I wish to thank Drs. M. Forbes, R. Hewitt, S. Kushner and J.P. Thomas for their invaluable assistance and suggestions in the preparation of this chapter.

Suggested Reading

Allison, R.F., *et al.* New Antituberculous Agents: Laboratory and Clinical Studies. *Annals of the New York Academy of Science.* 1966, **135**, 681-1120.

Doster, B., Murray, F.J., Newman, R., & Woolpert, H.F., Ethambutol in the Initial Treatment of Pulmonary Tuberculosis; U.S. Public Health Service Tuberculosis Therapy Trials. *American Review of Respiratory Diseases*, 1973, **107**, 177-190.

Freerksen, E., Orlowski, E.H., & Thumin, J.H. Experimental and Clinical Evaluation of the Tuberculostatics Capreomycin, Isoxyl, Myambutol, Rifampicin. *Antibiotics et Chemotherapia*, New York: Munchen, 1970, **16**, 536.

Hellstrom, P.E., Repo, U.K., & Mattson, K. New Drugs in Tuberculosis. *Drugs*, 1971, **1**, 349-353.

Horowitz, O. Public Health Aspects of Relapsing Tuberculosis. *American Review of Respiratory Diseases*, 1969, **99**, 183-193.

Johnston, R.F. & Wildrick, K.H. "State of the art" review. The impact of chemotherapy on the care of patients with tuberculosis. *American Review Respiratory Diseases*, 1974, **109**, 636-664.

Tao, J.C. Tuberculosis Control in the Western Pacific Region of WHO 1951-1970. *WHO Chronicle*, 1973, **27**, 507-515.

Thomas, J.P., Baughn, C.O., Wilkinson, R.G., & Shepherd, R.G. A new Synthetic Compound with Antituberculous Activity in Mice: Ethambutol [Dextro-2,2'-(Ethylenediimino)-di-l-butanol]. *American Review Respiratory Diseases*, 1961, **83**, 891-893.

Wilkinson, R.G., Shepherd, R.G., Thomas, J.P., & Baughn, C. Stereo-Specificity in a New Type of Synthetic Antituberculous Agent. *Journal of the American Chemical Society*, 1961, **83**, 2212-2213.

Youmans, G.P. & Youmans, A.S. Experimental chemotherapy of tuberculosis and other mycobacterial infection. *Experimental Chemotherapy*, New York: 1964, Vol. II: 393'499.

Tuberculosis Mortality in the United States, Canada and Western Europe. *Statistical Bulletin* (Metropolitan Life), 1971, **52**, 2-5.

CHAPTER 5

Can We Make It Useful? The Novel Therapeutic Approach

Irwin J. Pachter

Scientists devoted to the development of new drugs frequently encounter compounds with unusual biological effects. But can these effects be converted to practical utility, to new modes of therapy? Such factors as acceptance by the patient and society, separation of useful and toxic effects, route of administration and duration of action, cost and availability all need to be worked out before utility is achieved.

The initial discovery frequently becomes common knowledge many years before practical results. To the majority who are aware of the discovery, the obstacles to utility may appear insurmountable. The biological acitivity may be fleeting when long duration of action is required. The cost of production may be overwhelming and far beyond what the majority of patients can afford. Side effects may be life-threatening and little hope may exist that useful and toxic properties can be separated. Scientists may recognize the futility of trying to make a useful therapeutic agent of an initially synthesized member of a new biological series.

But then a single individual, or a small group coming together perhaps by chance, will be convinced that a useful product can be developed. Conviction, the spur to imagination, may carry the effort through to a useful conclusion. A program may be organized to overcome the obstacles by synthesis of new but related molecules with fewer drawbacks. Through careful examination of structural features, ways may be found to prepare analogous compounds with more prolonged action. Hints of fewer side effects in certain members of the group may suggest molecular changes that could lead to greater safety. Others, seeing the result, will then think, "How simple, why didn't we think of that?" On the other hand, the obstacles may prove truly insurmountable, at least for the moment, and failure may result. Others will think, "They wasted their time; I knew all along it couldn't work."

In almost every pharmaceutical firm today there are many such development efforts in progress. A good number will fail. Some will succeed. I would like to describe, as an ongoing example, attempts in progress to relieve pain and combat narcotic addiction and drug abuse through use of narcotic antagonists.

The first narcotic antagonist was synthesized by the German chemist von Braun in 1914. The following year his biologist collaborator, Julius Pohl, published a paper showing that these compounds can counteract the effects of morphine and codeine.

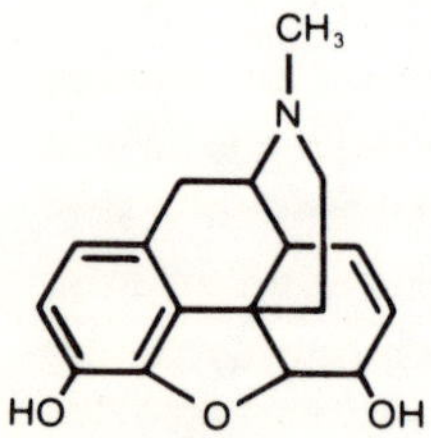

Morphine

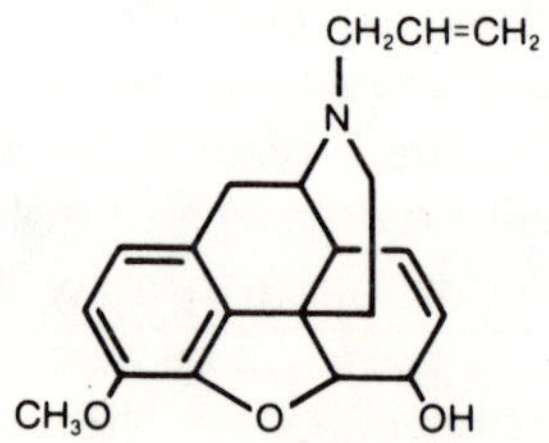

N-Allylnorcodeine

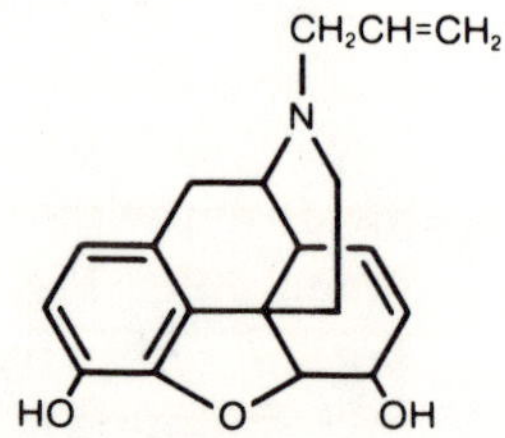

Nalorphine

N-allylnorcodeine, the first of the narcotic antagonists, was of potential use in counteracting the side effects of morphine overdosage, such as respiratory depression and nausea, but it was not sufficiently potent. Around 1940, the stronger narcotic antagonist, nalorphine, a close relative to N-allylnorcodeine, was prepared. It was made available to the medical profession by Merck & Co.

Narcotic antagonists such as nalorphine were known to counteract almost all of the effects of morphine and other potent narcotics, but in the early 1950s clinicians Louis Lasagna and Henry Beecher at Harvard Medical School asked some questions that were still unresolved. Can a morphine-nalorphine mixture be devised that will not have the principal toxic effects of morphine (respiratory depression, nausea and vomiting)? Does nalorphine counteract the pain-relieving power of morphine or just its side effects? To answer these questions it was necessary to make a detailed study of nalorphine itself in human trials. Laboratory scientists had previously failed to produce significant analgesia with nalorphine in animals. It was, therefore, a surprise when nalorphine was found to have powerful analgetic action in man. Paradoxically, nalorphine, which counteracts respiratory depression induced by morphine overdoses, was found to induce respiratory depression itself. Furthermore, mixtures of nalorphine with morphine produced side effects similar to those obtained with morphine itself. It was unfortunate that the analgesia induced by nalorphine, a nonaddicting drug, was accompanied by a new spectrum of side effects including unpleasant feelings of disorientation and hallucinations. Nalorphine could not be used for purposes other than the treatment of narcotic overdosage.

It became apparent to a few, however, that a new approach had been found to a long-sought goal—the development of a potent nonaddicting analgetic agent. The search began for the analgetic narcotic antagonist that would not provide unpleasant side effects at therapeutic pain-relieving doses. At the forefront of this research was a team at Winthrop Laboratories led by Sydney Archer and Louis Harris and another at the University of Rochester led by Marshall Gates who had previously supervised the first laboratory synthesis of morphine. The researchers realized that available animal models for pain, in which nalorphine produced no analgesia, could not be used to find a better drug of its class. It also became obvious to them that the old tests for analgesia in animals had led invariably to agents that ultimately proved addictive in man. Scientists devised new animal tests that showed the analgesia of nalorphine and were hopefully predictive of nonaddicting analgesia in man. New narcotic antagonists were synthesized and tested by the new procedures and a few were selected for human trial.

Those evaluated in man for analgesia all appeared to have some potential for hallucinogenic action—some more, some less. Generally, it was found that the most potent antagonists, such as cyclazocine from Winthrop and cyclorphan from the University of Rochester, were the most hallucinogenic. Pentazocine, a weak antagonist, with only slight potential for abuse and with relatively rare hallucinogenic action, was developed for the relief of pain by Winthrop Laboratories.

Then naloxone was synthesized in the private laboratory of Mozes Judah Lewenstein, a refugee who fled to the United States to escape Nazi aggression. The properties of naloxone were estraordinary. One part of the compound was able to block the effects of 100 parts of heroin. It produced no hallucinations, even at very high doses. In fact, it was even able to counteract and reverse hallucinations produced by all the other narcotic antagonists. Naloxone abolished respiratory depression, nausea, constipation, convulsions and all other effects produced by narcotics and narcotic antagonists, including their analgetic properties. Naloxone was a "pure" antagonist. It did not relieve pain, and it had no potential for addiction. Naloxone was developed during the 1960s by Endo Laboratories, a division of the du Pont Company and introduced commercially in 1971 as a narcotic antidote.

Because naloxone can block the action of all narcotic drugs of the morphine-heroin class, there were some scientists, including William Martin and Donald Jasinski of the United States Addiction Research Center, who thought that naloxone might offer an alternative to addicting methadone in the treatment of narcotic addicts. Methadone in place of heroin substitutes a legal narcotic for an illegal one. The patient remains an addict to narcotics. Naloxone, on the other hand, is nonaddicting and would serve to prevent addiction by blocking the action of narcotic drugs of abuse. A former addict who had taken naloxone, according to the theory, would feel no satisfaction from the injection of heroin since the narcotic effects would be blocked by naloxone. Patients who were formerly addicts were treated with naloxone by Max Fink of New York Medical College and then given doses of heroin. They felt no effect. The theory worked but naloxone was not a useful agent for the purpose. Although extremely potent by injection, naloxone is short-acting, and its blocking effects are usually gone after two to three hours. Oral-

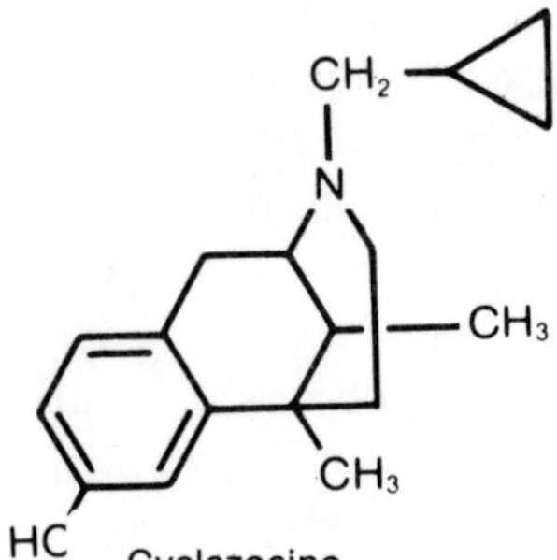

Cyclazocine

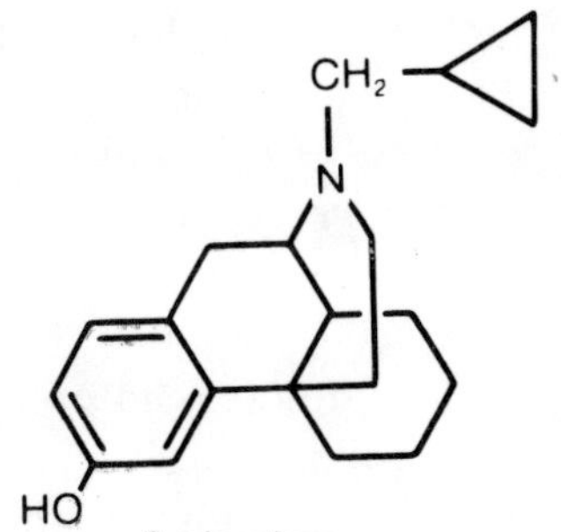

Cyclorphan

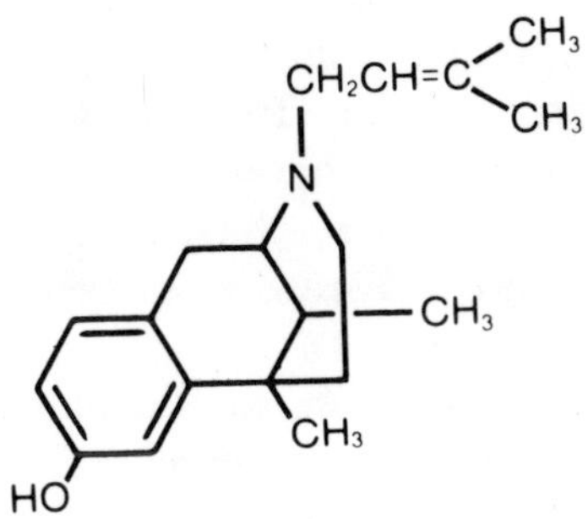

Pentazocine

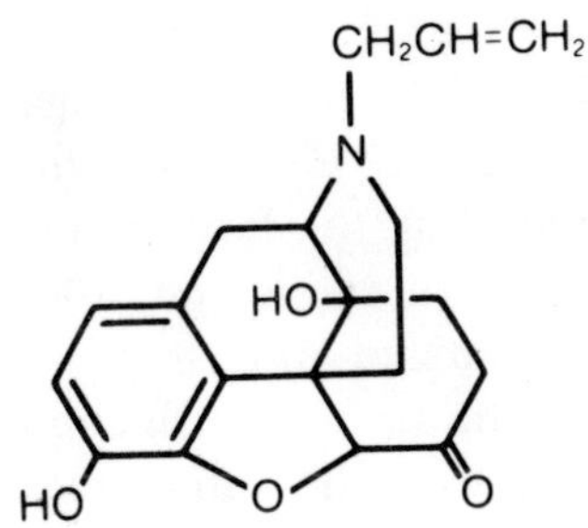

Naloxone

ly, it is rapidly metabolized, and in order to protect an addict from heroin challenge for 24 hours, huge oral doses of one to three grams per day are required. At these dose levels naloxone would be extremely expensive and in short supply.

On the other hand, because of its unusual properties, naloxone may be the ideal agent for combating the abuse of legitimate drugs. When very small amounts of naloxone are added to these drugs and the resulting combinations are taken *orally* in prescribed dosage, the naloxone is not effective and desired therapeutic results are achieved. When the drugs are processed for attempted abuse *by injection*, however, the naloxone immediately works to block narcotic effects and, in addicts, symptoms of withdrawal, including nausea, vomiting, sweating and general malaise, are precipitated.

Experiments along these lines with naloxone combinations have been conducted by Bristol Laboratories and by the United

States Addiction Research Center at Lexington, Kentucky.

Paregoric is one example of a drug whose abuse can be prevented in this way. Paregoric is a camphorated tincture of opium. Over 13 million ounces are sold in the United States each year for the control of diarrhea. In the *Federal Register*, Vol. 36, No. 203 (October 20, 1971) it is stated that "abuse of paregoric by addicts who process it into a form for intravenous administration is well-known and well-documented in medical literature . . . it is in the public interest for paregoric to be restricted to prescription sale."

Animal experiments at Bristol Laboratories show that such abuse of paregoric can be eliminated entirely by adding a mere 0.3 mg of naloxone to each fluid ounce of the drug. When boiled down, filtered and injected into rats, naloxone-treated paregoric is dramatically different from untreated paregoric. Rats receiving the former showed no change in their behavior while those receiving the latter rapidly became prostrate in a deeply narcotized state.

The discovery that potent narcotic antagonists such as naloxone can be almost totally free of side effects led researchers at Endo Laboratories to synthesize related compounds derived from opium. The goal was to prepare nonaddicting analgetics with fewer undesirable properties and orally effective narcotic antagonists with longer duration of action. Nalbuphine and naltrexone were derived. Both are presently undergoing clinical evaluation. The former is more potent than morphine with low potential for abuse. The latter is capable of blocking the action of heroin and other narcotics and is both longer acting and more potent than naloxone; it shows promise as a replacement for methadone in many patients.

These developments were noted at Bristol Laboratories where the theory was evolved that the relative freedom from undesirable side effects manifested by naloxone, nalbuphine and naltrexone was associated with the presence of the hydroxy group in the vicinity of the nitrogen atom.

Bristol then set out to prepare totally synthetic drugs designed to contain hydroxy or other oxygen-containing functions in appropriate positions. Total synthesis was deemed a

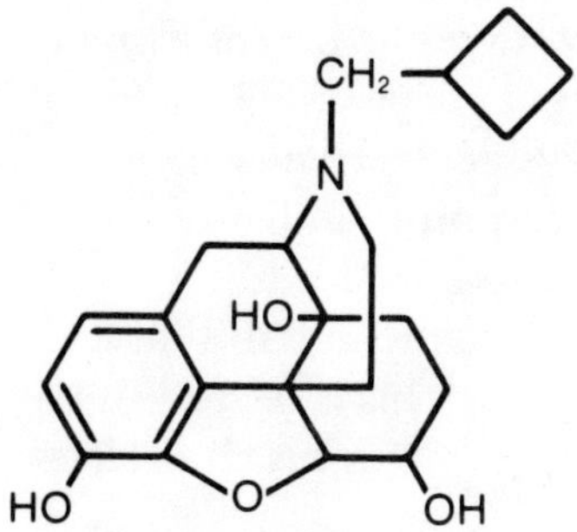

Nalbuphine

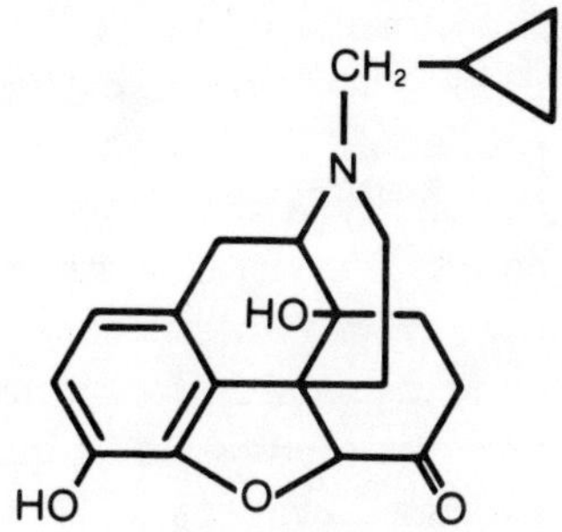

Naltrexone

worthy objective in order to eliminate our need to grow or import opium. Because total synthesis of these complex molecules was likely to be very expensive, it was realized at the outset of this work that the final products would have to possess potency many times those of morphine and codeine.

With previously synthesized cyclazocine and cyclorphan as models, Bristol synthesized butorphanol and its relative oxilorphan. Animal and human studies confirmed that introduction of the oxygen function led to a marked reduction in side effects while preserving the non-narcotic nature of the molecules.

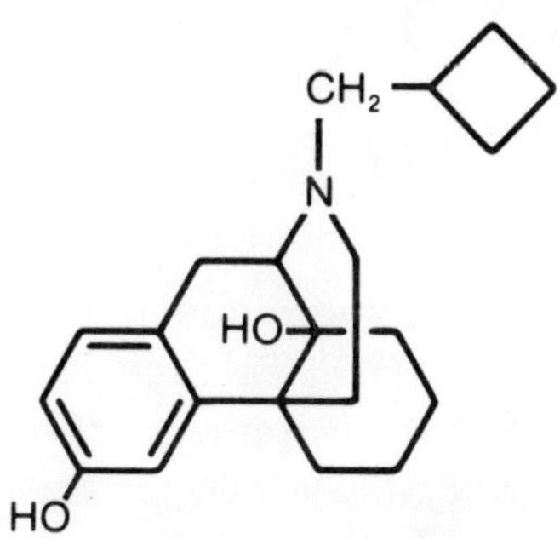

Butorphanol

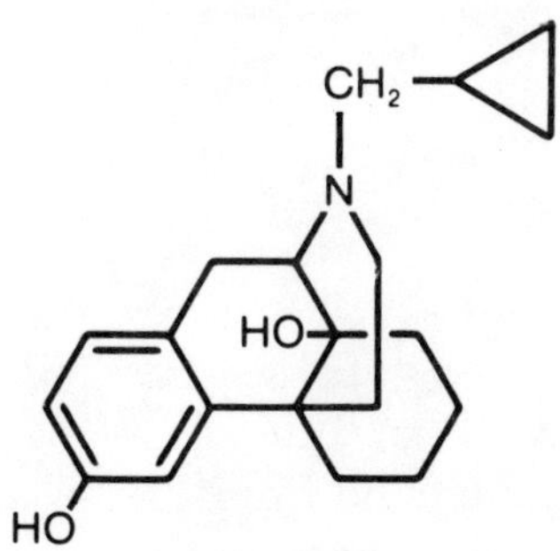

Oxilorphan

Butorphanol proved to be five to ten times as potent as morphine and to be an extremely safe drug. It appears incapable of producing life-threatening respiratory depression which occasionally accompanies the use of morphine and other narcotics. It is virtually free of the types of side effects observed with the earlier narcotic antagonist analgetics. Only in habitual narcotic drug users does it produce discomfort. This is because, as a narcotic antagonist, it induces abstinence symptoms. It is effective both orally and by injection.

Oxilorphan, which has little analgetic potency, is a very strong narcotic antagonist and is orally active at doses lower than those required for naltrexone or naloxone. Although its side effect potential is markedly less than that of cyclazocine or cyclorphan, it is greater than that of naltrexone, and exploration for still better molecules was indicated.

Encouraged by these results, Bristol synthesized two new compounds which have not yet been named but are designated by code numbers BC-2910 and BL-4566. These compounds, closer in structure to cyclazocine than to cyclorphan, also bear appropriate oxygen functions.

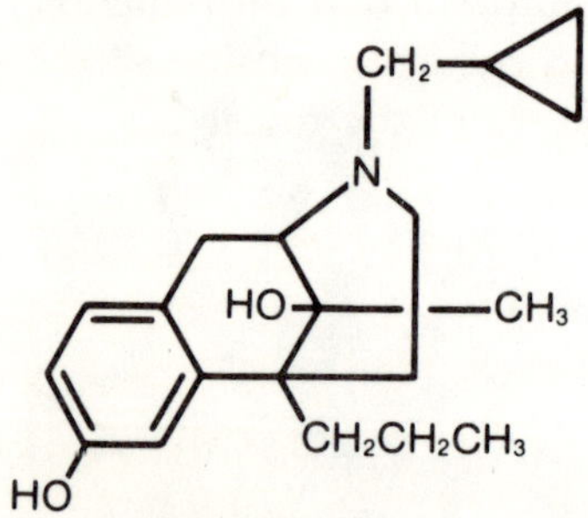

BC-2910

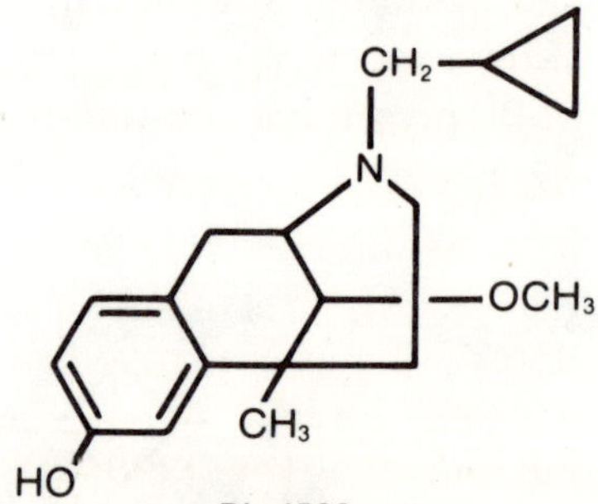

BL-4566

BC-2910 is a new pure antagonist that is more potent and longer acting than either naltrexone or oxilorphan. The National Institute of Mental Health received supplies of the compound from Bristol and began toxicological studies in animals in anticipation of human trials.

BL-4566 is a new narcotic antagonist analgetic with both the antagonist and analgetic potency of cyclazocine but without the latter's disturbing side effects. Parenteral doses 20 to 40 times the analgetic dose were administered to human volunteers without precipitating hallucinogenic or dysphoric reactions.

We have thus seen how narcotic antagonists, first synthesized in 1914, remained scientific curiosities until one was developed as a narcotic antidote in 1940. Better narcotic antagonist antidotes were subsequently developed. Drugs of this type, although life-saving in special situations, have minor commercial appeal and were developed by pharmaceutical companies predominantly as public service items.

Work with the antagonists led to the fortuitous discovery that they could be analgetic products with little potential for abuse. Important agents of this type are totally synthetic and promise to reduce or eliminate our dependence on the cultivation and importation of opium.

Narcotic antagonists have found a special use when mixed with narcotic drugs to eliminate injectable abuse by addicts.

Finally, narcotic antagonists are being developed in the hope that they may prove to be non-narcotic alternatives to treatment with methadone. This, again, is an area of little commercial appeal but of substantial potential importance to patients and to the public. Problems remain, however. For example, unlike methadone, the nonaddicting antagonists do not provide the detoxified addict with drug-like satisfaction. He is blocked from heroin effects, but how do we provide him with the incentive to keep returning to a treatment center for daily antagonist doses? The answer may lie in new methodology which employs a slowly released, injected form of the narcotic antagonist. It is expected that preparations will soon be available which will afford one to two months of protection against narcotics on single injection.

Will such drugs be acceptable? Is it proper to seek their routine use in areas of high narcotic addiction? What are the moral and legal issues? Will their use simply result in the more extensive abuse of non-narcotic drugs?

Obviously such questions cannot be answered by the pharmaceutical industry alone. If we know how to offer new therapy that may be useful in combating narcotic abuse we have an obligation to make it available. It will be for physicians to help determine whether the new methods and approaches will prove feasible. It will be for legislators to help determine how such products may be applied without restricting human rights.

Suggested Reading

Braude, M.C., Harris, L.S., Smith, J.P. & Villareal, J.E. Narcotic Antagonists. *Advances in Biochemical Psychopharmacology*. Vol. 8, New York: Raven Press, 1973.

deStevens, G. Analgetics. *Medicinal Chemistry*. Vol. 5, New York: Academic Press, 1965.

CHAPTER 6

Bioavailability and the Design of Drug Delivery Systems

Louis C. Schroeter

A revolution has taken place during the past decade in conceptual approaches to the design of drug products. The major thrust of this effort has been initiated and developed by the pharmaceutical industry. This is not to say that there have been no significant contributions from academic laboratories—far from it: the theoretical structures of the new pharmaceutical sciences have been greatly enlarged and strengthened by the prompt and extensive publication of imaginative research performed in academic, industrial and governmental laboratories. However, putting theory into practice has been virtually the exclusive domain of industry's scientists. Their efforts have translated new design concepts into new and improved drug products. Since the vast majority (well over 98 percent) of the prescriptions written in the United States call for manufactured drug products, the practical impact of improved product designs has been enormous.

Drug product designers have generally been trained in pharmaceutics. This field of study traditionally emphasized a detailed, though somewhat descriptive, knowledge of the physical and chemical properties of drugs and their compatability with the additives needed for preparing them in forms suitable for administering accurate doses to patients. Over the years, designers became increasingly concerned with maintaining full potency over the labelled shelf life of products. In the 1950s the principles of physical chemistry were utilized to improve quality. Such dramatic results were obtained that the field of pharmaceutics expanded to include the more rigorous mathematical and physicochemical approach toward drug product design known today as physical pharmacy. Designers were provided with powerful, theoretical tools. At the same time, drug systems were becoming more complex and requirements for stability more stringent. A further complication to the designer's task was added when patients began to reject the old idea that good medicine should taste bad. Ironically, this changing attitude on the part of patients seemed to coincide with the discovery of many potent drugs, all of which seemed to possess the worst possible bitter tastes. Thus, the designer became responsible for developing a drug product that would not only maintain its potency over the full life of the product but also maintain its appeal to the patient from the standpoint of color, flavor and odor. Balancing a variety of conflicting factors provided a spirit of excitement and challenge to the drug product designer in the late 1950s. This quality is still characteristic of the present-day designer's task and is often cited as one of the most rewarding aspects of his job.

From the earliest times, it had been known that the quality of the drug product affected its performance in man. As early as 1886, the Upjohn Company had marketed a specially made pill which, unlike others of its day, could be crushed to powder under a person's thumb. When taken, these friable pills rapidly released their drug content whereas some of the other pills of the day would pass through the body unabsorbed. Later, when compressed tablets became popular, drug product designers developed special laboratory tests to ensure that the tablets they produced would disintegrate. It was recognized that tablet disintegration into granules was a necessary prelude if the drug in the tablet was to be absorbed by the patient. Nonetheless, until the 1960s it was generally believed by medical scientists, and even by a good many drug product designers, that the design only rarely affected the drug's performance in man.

A new dimension to drug product design began when pharmaceutical scientists and others became fully aware of the broad range of therapeutically significant effects attributable to the design or to the method of producing new products. In 1961, John G. Wagner, then of The Upjohn Company Research Laboratories, published in the Journal of Pharmaceutical Sciences a bold and imaginative review article titled "Biopharmaceutics: Absorption Aspects," which summarized much of what was then known about the influence of formulations on the therapeutic activity of drug products. The term "bioharmaceutics" gained almost immediate acceptance for the many active programs dealing with formulation effects which were under way at that time in other pharmaceutical laboratories. Soon the literature was filled with research reports dealing with biopharmaceutical studies describing the relationship between the physicochemical properties of drugs in certain formulations and the biological effects observed following administration of these formulations. The most important and lasting result of the early work in the field now known as biopharmaceutics is that it has changed our way of thinking about the role of the designer of drug products and the information we must have to design optimum performance into the product.

In order to ensure that the dose of drug put into the product is "available" to the patient for its intended effect, the designer must consider how the drug should be used to treat the disease in the most effective way. The central concern of the designer must be to provide effective and reliable therapeutic agents. Information needed by the designer includes a detailed knowledge of the relationship between the amount of drug administered and the resultant biological effect or pharmacologic activity. In addition to the dynamic aspects of drug effect, it is essential that the designer have at hand information about the rate and extent of the drug's *absorption, distribution, metabolism* and *excretion* (ADME) following drug administration by the oral and parenteral routes. Characterization of drug action in this manner is described as a pharmacokinetic or ADME study. Finally, the designer must focus attention on the traditional concerns of pharmaceutics. The drug, and the materials with which it is to be combined, must be formulated into a system possessing optimum physical and chemical compatibility. Figure 1 serves to summarize what the new concept of drug product is all about.

Up to this point, we have mentioned the terms "drug" and "drug product" almost as if they could be used interchangeably. This is not so; it is important to make a clear distinction between the two: *a drug is the active component contained in*

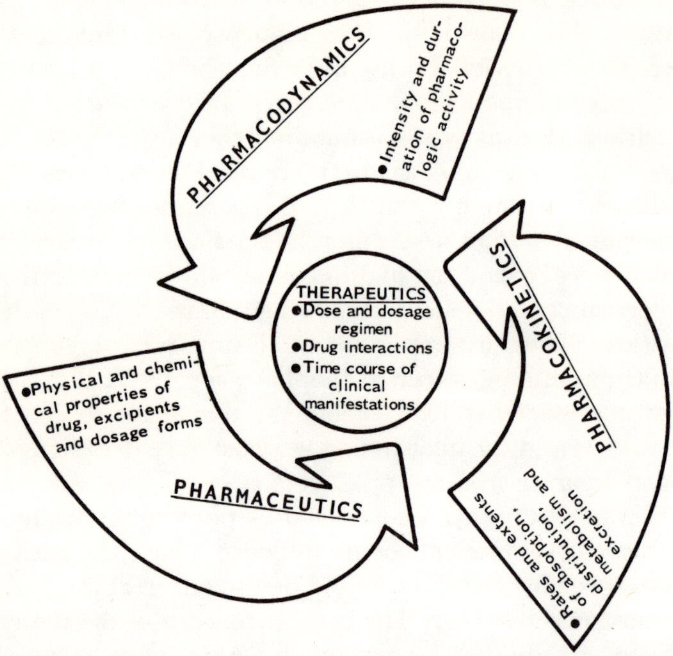

Figure 1. The new concept of drug product design requires new kinds of data.

what is usually a multicomponent drug product. Generally, *we speak of the drug product as being a dosage form.* This distinction is illustrated in Figure 2. A drug is a pure chemical compound; a dosage form is the physical form in which the patient takes the drug—it may contain one or more drugs along with other ingredients. The relationship between drug and dosage form may be represented as

$$\text{Drug} + \text{X} = \text{Dosage Form}$$

Where the X represents ingredient(s) fulfilling one or more of the following functions in the system:

diluent
buffer
surfactant
flavor
antioxidant
preservative
disintegrant
de-aggregant
wetting agent
suspending agent
emulsifier

The list is by no means complete—it is cited here to illustrate the potential complexity of drug products from a standpoint of components. The dosage form system may be simple (drug plus one or two X-type ingredients) or complex (drug plus many X-type ingredients); however, the system must always meet the test of providing *reliable* and *reproducible drug bioavailability*. This means that each individual dosage form must provide the optimum release of its drug content at the site of absorption in the body. For example, each capsule of an antibiotic drug product contains a measured dose of the antibiotic (drug) plus one or more X-type ingredients to ensure that the full dose of antibiotic is released from the capsule and optimally absorbed from

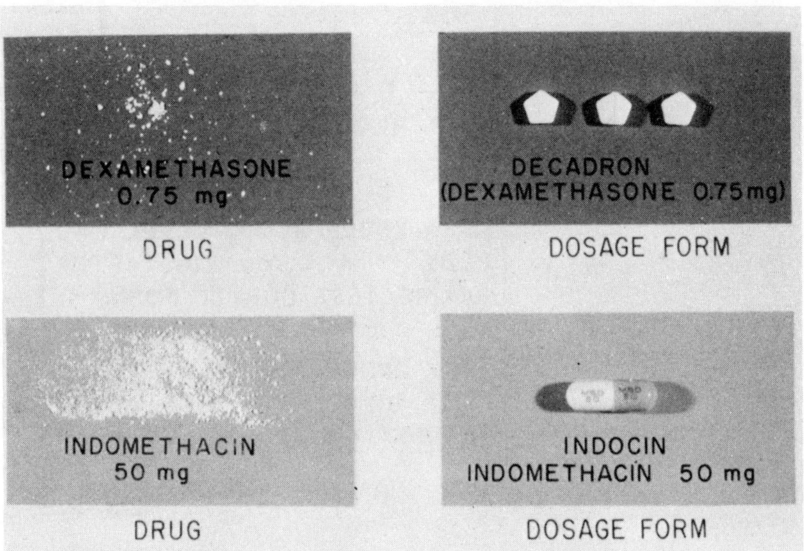

Figure 2. The distinction between drug and dosage form.

the patient's upper gastrointestinal tract. Each and every capsule must perform in exactly the same way, delivering its antibiotic content at the site of optimal absorption in each patient. It is in this sense that a dosage form functions as a drug delivery system.

Bioavailability is a measure of *how much* of the drug is absorbed or made available to the patient from a dosage form in a given time period, and how fast this process occurs. We have several ways of describing bioavailability—it may be represented as a ratio or as a percentage of the areas under drug blood level-time plots (see Fig. 3). Note that this is a comparison between the area obtained from the same dose of drug administered in a test dosage form and in a standard dosage form (generally a solution). It is implied that the comparison be made under crossover conditions, that is, that the same subject or group of subjects take both dosage forms. Ideally, the blood level should be followed for a sufficiently long period after each dose administration to the point when no significant amounts of drug remain in the blood.

BIOAVAILABILITY AS DETERMINED BY AREA UNDER BLOOD LEVEL CURVE

$$\%\ \text{AVAILABILITY} = \frac{\left[\begin{array}{l}\text{AREA UNDER BLOOD LEVEL-TIME}\\ \text{PLOT IN A GIVEN TIME AFTER}\\ \text{TAKING TEST DOSAGE FORM.}\end{array}\right]}{\left[\begin{array}{l}\text{AREA UNDER BLOOD LEVEL-TIME}\\ \text{PLOT IN A SAME TIME AFTER}\\ \text{TAKING STANDARD DOSAGE FORM.}\end{array}\right]} \times 100$$

Figure 3.

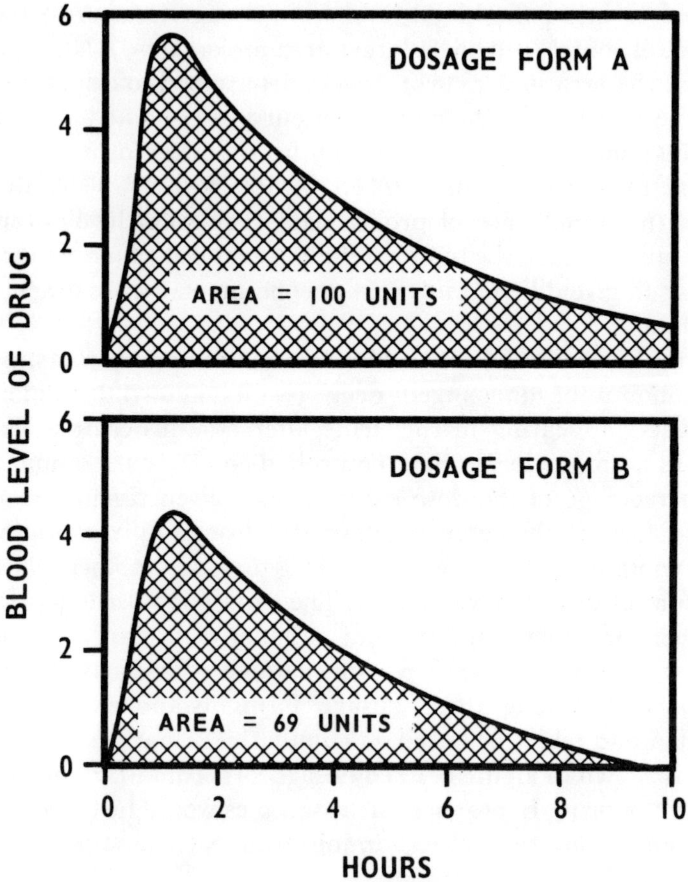

COMPARISON OF AREAS UNDER BLOOD LEVEL-TIME PLOTS OF IDENTICAL DOSES OF DRUG ADMINISTERED IN TWO DIFFERENT DOSAGE FORMS.

Figure 4.

A comparison of drug blood levels from identical doses (same amount of drug) given in two dosage forms is shown in Figure 4. In this example, based on 10 hours of sampling, the drug bioavailability from dosage form B would be 69%, assuming dosage form A as the standard or reference preparation. Carrying out the sampling for, say, 16 hours or whatever period that would be needed to bring the blood level from dosage form A to zero, would, of course, provide a larger area under curve A. If, for example, the area under curve A were determined to be 110 units after 16 hours while the area under curve B remained at 69 units, the relative drug bioavailability from dosage form B would appear to be decreased: (69/110) X 100 = 62.7%. This illustrates the importance of proper data evaluation in bioavailability studies; even the most carefully controlled experiment can appear to give different answers depending on the treatment of the data.

Drug bioavailability may also be determined by measuring the amount of unchanged drug (or its known breakdown products) appearing in the urine after administering a fixed dose in a dosage form under controlled conditions. Comparing the percentage of the dose excreted in a given time from one dosage form to the percentage excreted from a different dosage form containing the same dose of drug provides another relative measure of drug bioavailability. The expression used to calculate percent bioavailability from cumulative urinary excretion data is shown in Figure 5. An example of a bioavailability comparison among three dosage forms (standard solution, tablet A and tablet B) based on cumulative urinary excretion of drug is shown in Figure 6. Each dosage form contained the same amount of drug. In practice, urine samples would be taken for a sufficiently long time after administering each dosage form to

BIOAVAILABILITY AS DETERMINED BY CUMULATIVE URINARY EXCRETION MEASUREMENTS OF DRUG OR METABOLITE.

$$\%\ \text{AVAILABILITY} = \frac{\left[\begin{array}{l}\text{\% OF DOSE EXCRETED IN URINE}\\ \text{IN A GIVEN TIME AFTER TAKING}\\ \text{TEST DOSAGE FORM.}\end{array}\right]}{\left[\begin{array}{l}\text{\% OF DOSE EXCRETED IN SAME}\\ \text{TIME AFTER TAKING STANDARD}\\ \text{DOSAGE FORM.}\end{array}\right]} \times 100$$

Figure 5.

ensure that no additional quantities of drug or metabolite would be excreted. This point can be detected when the cumulation plot shows no further upswing—the top part of the S-shaped curve becomes parallel to the X-axis as shown for tablet B in Figure 6. Comparing the cumulative excretion of drug from tablet B to the solution after 10 hours gives the relative bioavailability of tablet B as 50 percent. This assumes, of course, that no further excretion of drug takes place after the 10-hour period of urine collection. In a well-designed experiment, this assumption would be checked by taking urine samples at, say, 12, 14, 16, 20 and 24 hours. Note that the percent bioavailability of the dose in tablet A compared to the standard dosage form (solution) is about 95 percent. Comparing the bioavaila-

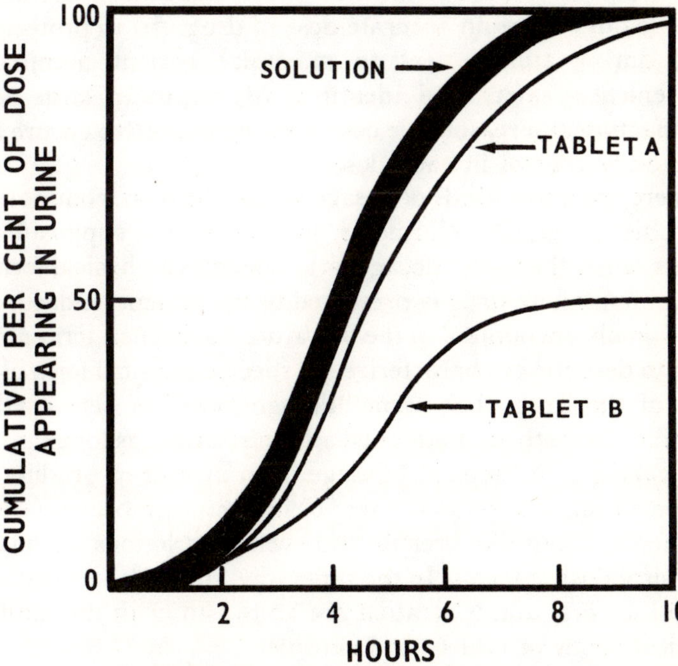

COMPARATIVE PLOTS OF URINARY EXCRETION OF DRUG WHEN ADMINISTERED IN SOLUTION AND IN TWO DIFFERENT TABLET FORMULATIONS.

Figure 6.

bility of tablet B to the solution we obtained the value of 50 percent; however, if we now make tablet A the reference dosage form, the bioavailability of tablet B becomes (50/95) X 100 = 52.6%. The bioavailability obtained from a solution of the drug usually serves as the ideal or target for the designer. However, a solution of the drug might not be acceptable because of inherent poor stability, bad taste or irritating properties. The designer must then maximize bioavailability by designing some other acceptable dosage form. Generally, this will be an oral solid dosage form. Designing such a form that will provide the same bioavailability as a solution is an almost impossible task (more about this later) so the designer usually strives for the best approximation permitted by the dosage form requirements. His objectives for design requirements of the dosage form are as follows: (a) to ensure an accurate dose of drug, (b) to protect the drug during storage, (c) to maximize patient acceptance (convenience, safety and identity), (d) to attain large scale manufacture at a reasonable cost, and (e) *above all* to ensure bioavailability control in each dose.

There are many kinds of dosage forms, the more common being tablets, capsules, ointments, injectables and suppositories. In this sense, the term "dosage form" denotes a physical form in which the dose of drug is presented to the patient. Indeed, one occasionally encounters in the literature the French term *presentation* to describe or characterize the specific physical form of the dose of the drug. The term "dosage form" is also used to describe the method of intended administration, as for example, in speaking of a "topical" dosage form as a drug product intended for application to the skin or mucous membrane. Greasy ointments, nongreasy creams and even fluid lotions all qualify as topical dosage forms. In the same way, solid dosage forms intended for oral administration are known to us in the familiar physical forms of tablets and capsules.

Special design problems are encountered with each kind of dosage form, but by far the greatest design challenge occurs with oral solid capsules and tablets. As one might expect, the dosage form most difficult to design properly (tablets) is the one most often prescribed! Tablets are greatly favored because they provide exact drug doses, protect the drug during storage and are convenient and easy to take. However, as depicted in Figure 7, tablets must first disintegrate and deaggregate before the solid drug particles can go into solution. This is an inherent disadvantage to the designer of a tablet because *drugs must be in solu-*

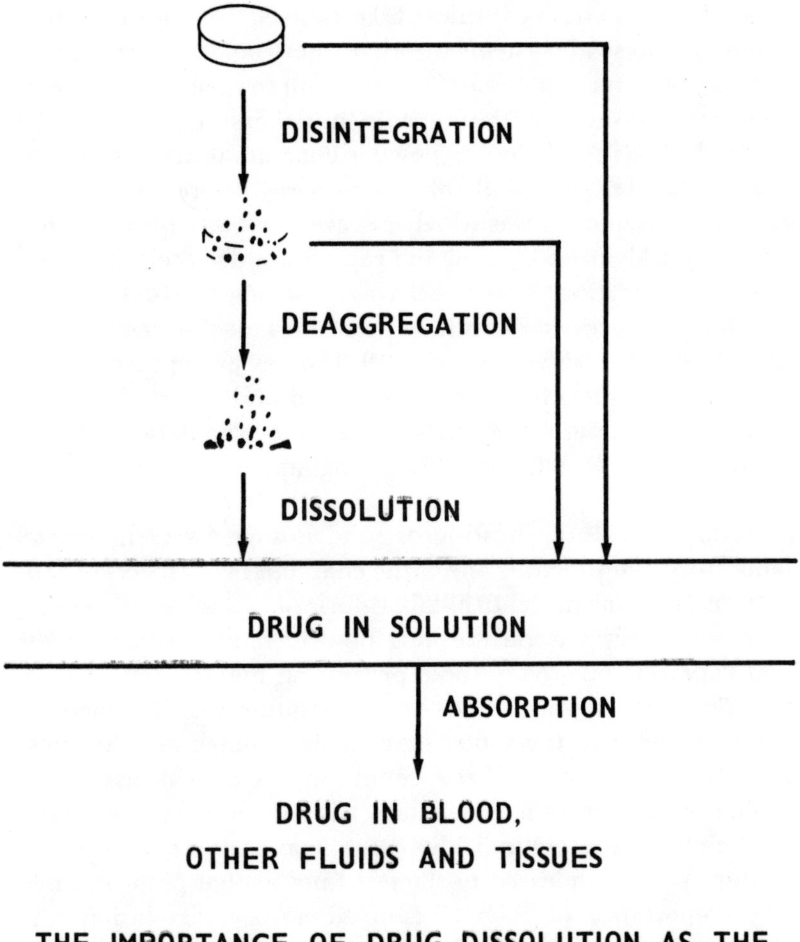

THE IMPORTANCE OF DRUG DISSOLUTION AS THE FIRST STEP IN THE PROCESS OF DRUG ABSORPTION DISTRIBUTION, METABOLISM, EXCRETION

Figure 7.

tion to be absorbed. Whatever hinders *any* of the steps of the dissolution process can potentially slow drug absorption in the body and, in so doing, decrease bioavailability of the drug contained in the tablet.

The requirement for drug dissolution from a tablet dosage form as a prerequisite to absorption in the body also applies to drugs formulated as capsules. Like tablets, capsules represent complex physical systems in that they normally contain a number of X-type ingredients in addition to their drug content. The gelatin shell of the capsule must first dissolve in the stomach fluids to expose its powder contents to the disintegration— deaggregation— dissolution process. Before the advent of biopharmaceutics, it was widely believed that all capsule formulations quickly disintegrated and rapidly yielded their drug content for absorption. This belief was so strong that designers of the day rarely tested capsule formulations for disintegration or dissolution characteristics. Since 1961, many examples of poorly dissolving capsules have been discovered and reported. Today, no reputable designer assumes his capsule formulation releases its drug content without conducting appropriate laboratory tests.

Today, dissolution testing of solid dosage forms in special laboratory equipment is one of the characteristic aspects of capsule design. Pharmaceutical laboratories always have had complex machinery for transforming powder mixtures into tablets and capsules, but today these processing machines exist side-by-side with sophisticated testing equipment designed to measure the way the tablet or capsule disintegrates and dissolves under a variety of test conditions. A typical dissolution profile of drug release from a tablet is shown in Figure 8. The solid circles represent individual assays or determinations of the amount of drug released to the test fluid at that point in time. The importance of such *in vitro* laboratory dissolution rate measurements is shown in Figure 9, which represents an idealized relationship between dissolution rate (expressed as time required for 90% of the drug to go into solution) and the percent bioavailability. At this point, the problem of maximizing drug bioavailability from a solid dosage form seems

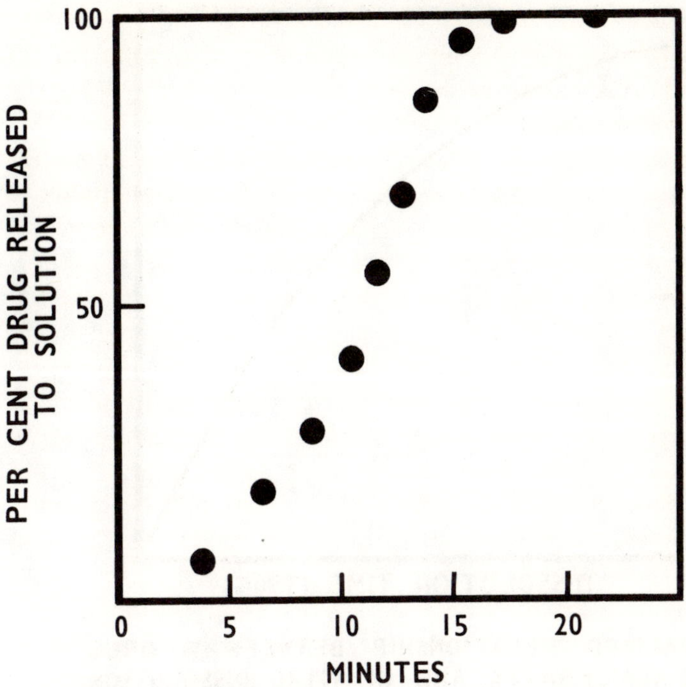

TABLET DISSOLUTION TEST – A PROFILE OF DRUG RELEASED TO SOLUTION AS A FUNCTION OF TIME.

Figure 8.

deceptively simple. One would expect that making the tablet dissolve as fast as possible in the laboratory dissolution test would ensure drug bioavailability. We shall see that not only is the relationship complex, but that the specific *in vitro* dissolution test must be validated for each drug.

The drug, the formulation and the process for making the dosage form may all influence the dissolution or release of the drug from the dosage form. In the case of oral solid dosage forms such as capsules and tablets, some of the many specific factors

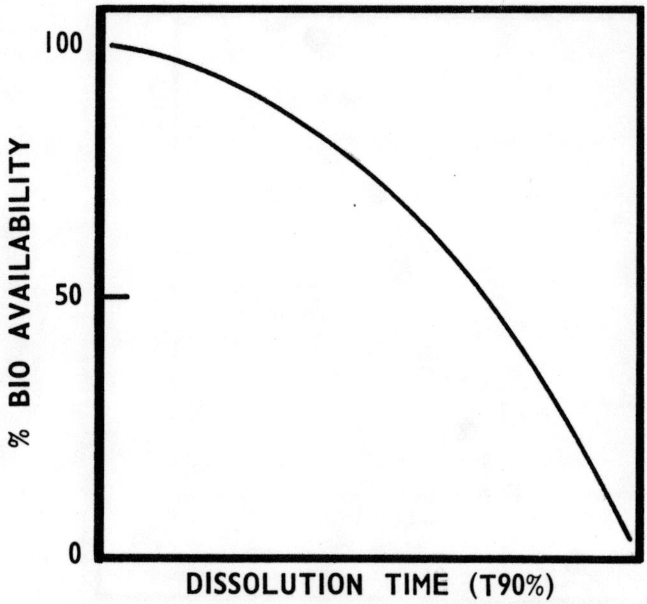

IDEALIZED RELATIONSHIP BETWEEN % DRUG BIO-AVAILABILITY AND *IN VITRO* DISSOLUTION RATE OF DRUG IN DOSAGE FORM WHEN T90% IS THE TIME FOR 90% OF THE DRUG TO GO INTO SOLUTION.

Figure 9.

now known to affect dissolution characteristics are as follows:

Drug	*Formulation*	*Process*
polymorphism	surfactant	method of granulation
solvation	disintegrant	compressional force
ionization	lubricant	speed of compression
particle size	wetting agents	method of mixing
surface area	buffers	mixing equipment
eutectics	diluent	time of mixing
solid solution	neutral salts	milling equipment
complexes	granule size	compacting equipment
	size distribution	drying conditions
	coating or shell composition	

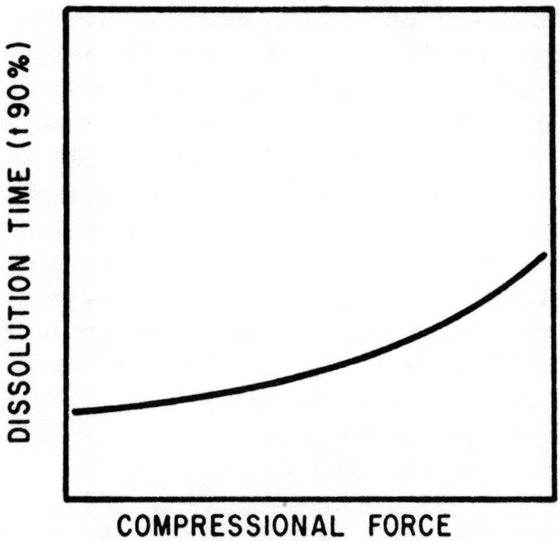

RELATIONSHIP BETWEEN COMPRESSIONAL FORCE REQUIRED TO FORM TABLET AND RESULTANT IN VITRO DISSOLUTION RATE OF TABLET. t90% REPRESENTS TIME REQUIRED FOR 90% OF DRUG TO GO INTO SOLUTION.

Figure 10.

Since drug dissolution from the design form is a prerequisite condition for drug absorption in the body, any of the factors shown above may alter bioavailability if the product is not properly formulated. If there was only a single relationship as represented in Figure 10, the designer's problem would be relatively simple. This figure shows the dissolution rate of the drug from a specific tablet as a function of the compressional force used to make the tablet. The higher the compressional force, the longer the time required for the tablet to release 90 percent of its drug content. Faced with this kind of information, the designer would try to compress the powder just hard enough to make the tablet hold together during packaging, shipment and storage without unduly slowing its drug releasing characteristics.

The relationships affecting drug release from a tablet are usually quite complex. As a somewhat simplified example of real-life complexities, we have represented the relationship between drug release from a tablet and each of three variables taken independently in Figure 11. By expressing each of the variables in terms of some arbitrary or standard value, on the horizontal axis, we can better visualize the relationships. At a constant composition of the powder formulation, increasing the pressure to produce the tablet (moving toward +1) has no effect on the time required for the tablet to release 90 percent of its drug content (t-90%) in the standard dissolution test. Similarly, as the pressure to form tablets from the powder is decreased (moving toward -1), there is no change in the t-90% value. This kind of formulation provides drug release independent of tabletting pressures in the range of the test processes. Changing the lubricant concentration in the specific powder formulation, while holding all other components constant and compressing the powder formulation into a tablet at constant pressures, results in a tablet which releases drug slower (higher t-90%) as more lubricant is added (moving toward +1 on the X-axis). In a similar fashion, when one holds the lubricant concentration constant, but varies the disintegrant concentration and then compresses the powders at constant pressure, the relationship of faster drug release (lower t-90%) is obtained with increasing disintegrant concentration (moving toward +1 on the X-axis). In these simplified examples, only one variable was changed at a time. However, in the real world of dosage form designing, there are usually a large number of variables and the product designer must determine the optimum composition and process conditions by balancing all factors. It is for this reason that the computer is being used with increasing frequency to design and evaluate experiments aimed toward selecting the best formulation and process.

The specificity of the relationship between an *in vitro* laboratory test of drug dissolution from a dosage form and drug bioavailability can be illustrated with the five experimental dosage forms of Merck's non-steroidal anti-inflammatory drug indomethacin (Indocin®). As shown in Figure 12, the times for

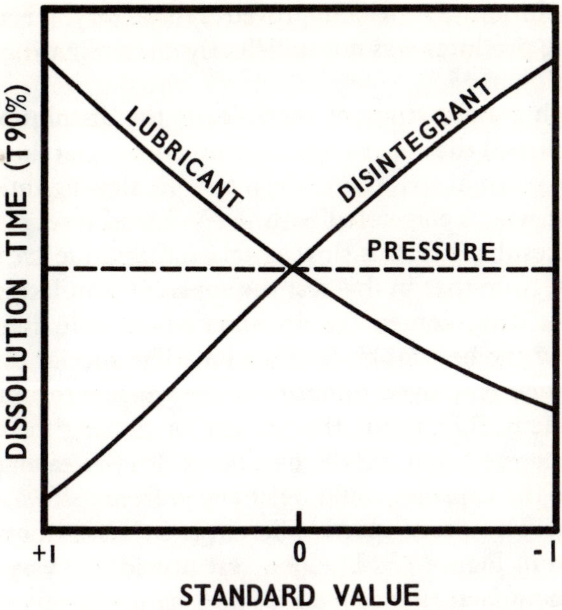

COMPLEX RELATIONSHIP BETWEEN FORMULATION AND PROCESSING VARIABLES FOR A SPECIFIC SOLID DOSAGE FORM. T90% REPRESENTS TIME REQUIRED FOR 90% OF DRUG TO GO INTO SOLUTION.

Figure 11.

90% of the drug to be released from the one capsule and four tablet dosage forms in this *in vitro* laboratory test, averaged around 6 to 9.5 minutes. Recalling the idealized relationship of Figure 9, one could easily conclude that all the experimental dosage forms of indomethacin would give the same blood level, since all exhibited essentially the same drug release characteristics in the laboratory dissolution test. However, when the dosage forms were tested in dogs, it was clear that there were significant differences in the resultant blood levels. As shown in Figure 13, the differences were especially striking in the two-hour period immediately following dose administration. The *in vitro* dissolution test which proved satisfactory for designing other drug products was not sufficiently discriminating for indomethacin!

Through careful study of variables of the dissolution test, it was established that the rate of agitation of the test fluid was too fast in the original test; therefore, a specific slow agitation dissolution test which correlated with *in vivo* blood level results was devised. Results from this slow agitation technique are shown in Figure 14. Note that in this test the capsule formulation yielded 90% of its drug content in 45 minutes—double the amount released by the best tablet formulation. The special dissolution test was then employed to optimize the capsule formulation of indomethacin. By altering the process for making the capsules, it was discovered that the dissolution or drug-releasing characteristics of the capsules could be changed from "slow" to "fast" dissolving. The pronounced differences in dissolution profiles are shown in Figure 15. The new test provided a way to select the best manufacturing process for making the fastest dissolving capsules.

Validation of the special slow agitation *in vitro* dissolution test

DISSOLUTION RATES OF INDOMETHACIN FROM DOSAGE FORMS USING THE U.S.P. APPARATUS FOR AGITATION*†

DOSAGE FORM	CHARACTERISTIC	TIME FOR 90% TO DISSOLVE (MIN.)	
CAPSULE	–	9.5	(7-15)
TABLET #1	METHOCEL GRANULATION	7.5	(7-10)
TABLET #2	CARBOWAX GRANULATION	7.5	(7-10)
TABLET #3	PVP GRANULATION	7.5	(7-10)
TABLET #4	ALCOHOL GRANULATION	6	(5-7)

*ALL VALUES REPRESENT AVERAGES OF AT LEAST SIX RUNS IN pH 7.2, 0.05 M., PHOSPHATE BUFFER AT 37° C.

†ALL DOSAGE FORMS MADE FROM FINELY MILLED SOLID, MMD-5 MICRONS.

Figure 12.

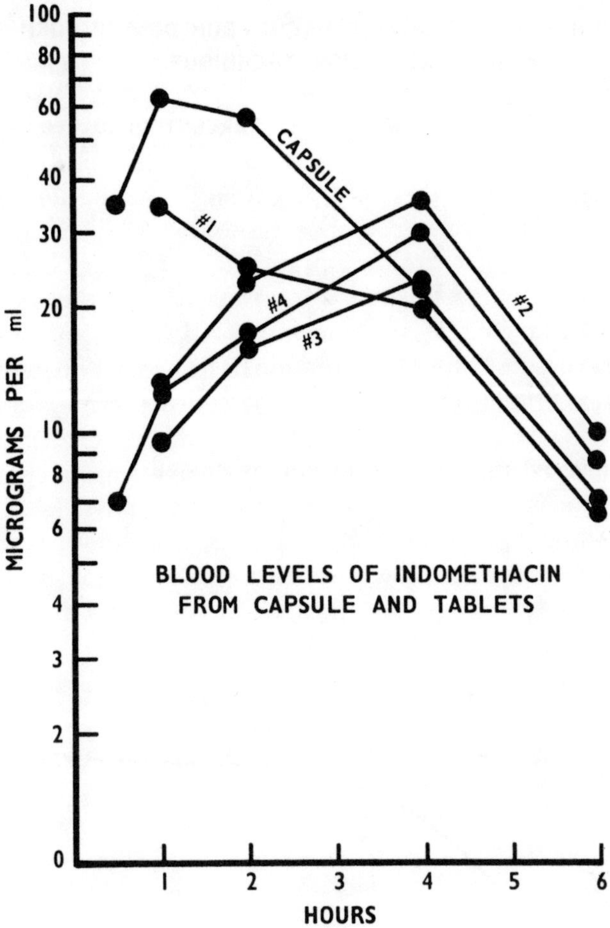

Figure 13.

procedure involved testing both slow-and-fast-dissolving capsules in a 12-patient crossover experiment. The peak plasma level obtained with the slow-dissolving capsules (Fig. 16) was just a bit above two micrograms per milliliter of plasma. The peak for the fast-dissolving capsules (Fig. 17) was about three micrograms per milliliter of plasma. Note also that the average peak plasma level from the fast-dissolving capsules occurs somewhat earlier than the peak produced by the slow-dissolving capsules.

DISSOLUTION RATES OF INDOMETHACIN FROM DOSAGE FORM USING A SLOW AGITATION TECHNIQUE

DOSAGE FORM	t_{90}–U.S.P.*	PERCENT DISSOLVED – BY SLOW AGITATION	
		IN 15 MIN.	IN 45 MIN.
CAPSULE	9.5	40	90
TABLET #1	7.5	43	43
TABLET #2	7.5	20	25
TABLET #3	7.5	3	3
TABLET #4	6	32	44

*CONDITIONS OF TEST AS PREVIOUSLY DESCRIBED.

Figure 14.

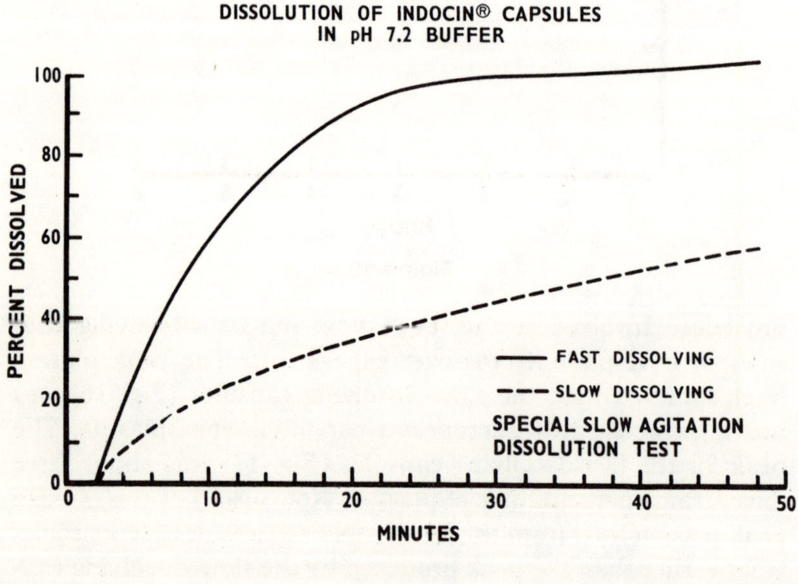

Figure 15.

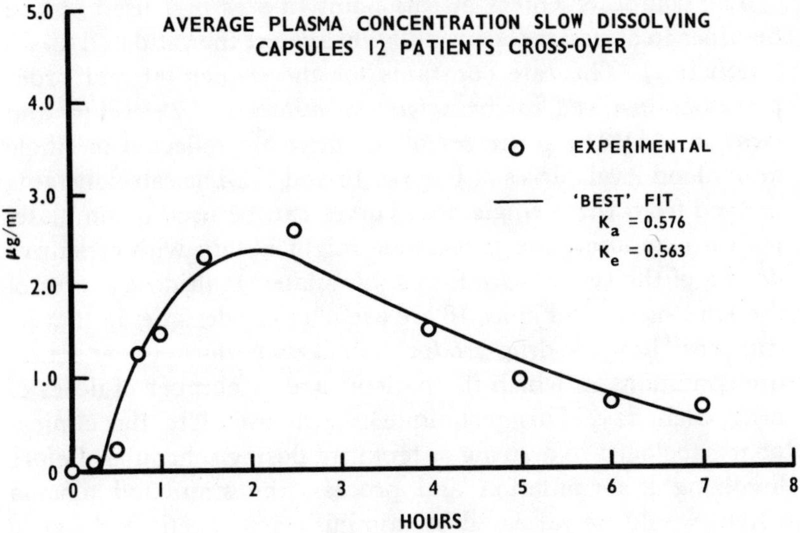

Figure 16.

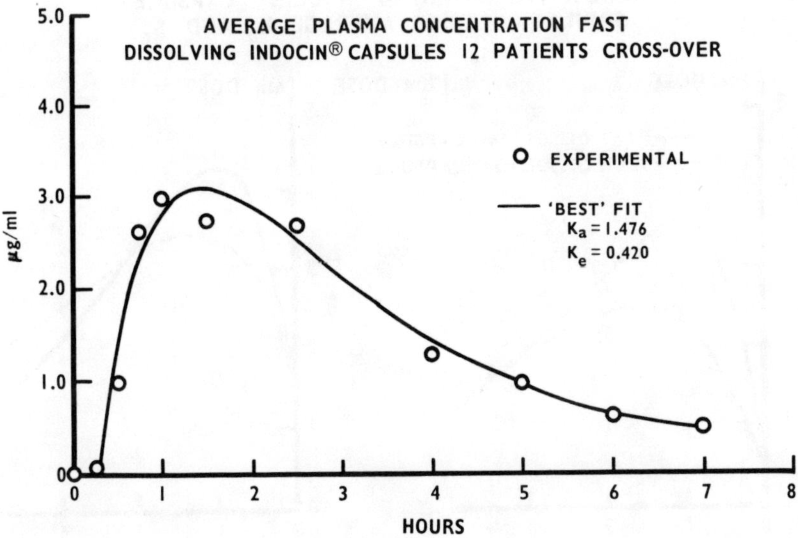

Figure 17.

The computer enters at this point (it was first used by the designer to optimize the formulation against the validated dissolution test). The rate constants for the sequential first order processes involved in the *absorption*, *distribution*, *metabolism* and *excretion* (ADME) processes of the drug are reflected in single dose blood level curves of Figures 16 and 17. The rate constants derived from these single-dose curves can be used to simulate, with a computer, the curves one might obtain with repetitive dosing of the two dosage forms. Simulated multidose curves of the kind shown in Figure 18 are useful to the designer in that he can "see" how the drug product will likely perform under actual use conditions in which the patient takes a number of doses of drug each day. This technique is also useful to the clinical pharmacologist in arriving at tentative dosing schedules. Before finalizing a formulation and process, the simulated plasma curves would be validated by administering repetitive doses to subjects and sampling at times determined from the curves. The simulated and actual curves do not always coincide. There may

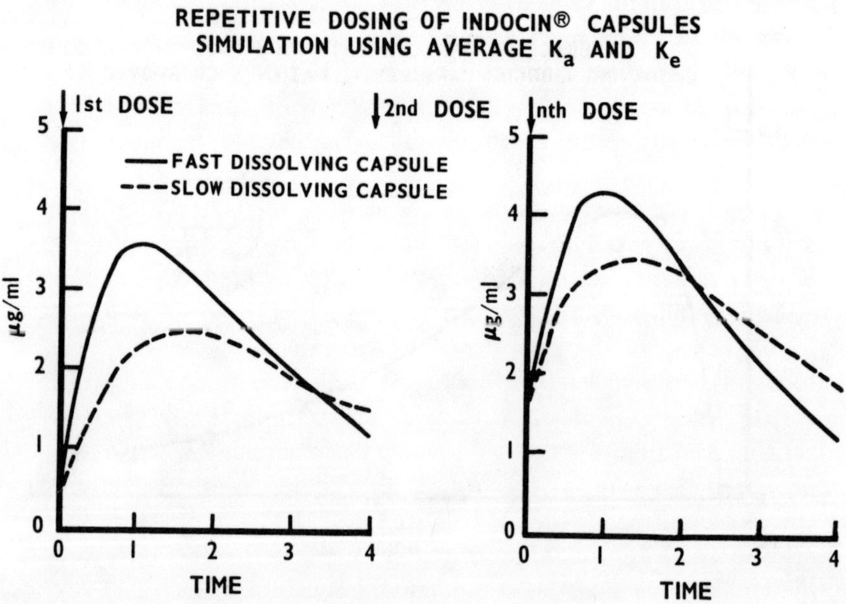

Figure 18.

be a mismatch if the assumption of first order kinetics for the absorption or elimination phase is not correct, or if drug absorption from the dosage form is affected by food or other factors not considered in the computer model. The difference between the simulated and actual plasmas curves provides useful information for the designer and for the use of the final product.

The last step in the design is to ensure that the dosage form made in full-scale production will provide the same reliable and reproducible blood levels as those which were first tested clinically. The designer establishes the initial specifications for an acceptable product on the basis of a validated dissolution test such as is shown in Figure 19. The test is said to be "validated" because the dissolution profile "meets specs" relates to capsules that were tested in man and found to be bioavailable, while those capsules giving unacceptable plasmas levels produced the lower dissolution profile marked "fails specs". Capsules made in production facilities utilize the special agitation dissolution test. Those batches with dissolution profiles like the upper curve "meets specs" would be passed by Control as being suitable for sale, whereas any batch with a profile like the lower curve "fails specs" would be rejected. Additional specifications based on other standardized dissoluation test procedures might be developed later for routine control of drug product quality but these specifications would be correlated with the initial tests.

Today, the pharmaceutical scientist in industry views himself not simply as a formulator of dosage forms but as a designer of drug delivery systems concerned with all design aspects that affect the bioavailability and performance of the drug product.

Awareness of the importance of formulation factors on drug bioavailability has had a profound influence on the design of conventional capsules and tablets. More importantly, the conceptual and methodological advances of the past few years have prepared the way for the design of what we might call "nonconventional" drug delivery systems. These are systems that release the drug to a target organ at an optimal rate. One example is a vaginal ring device which will continuously release very low levels of progesterone in the vagina. This system is capable

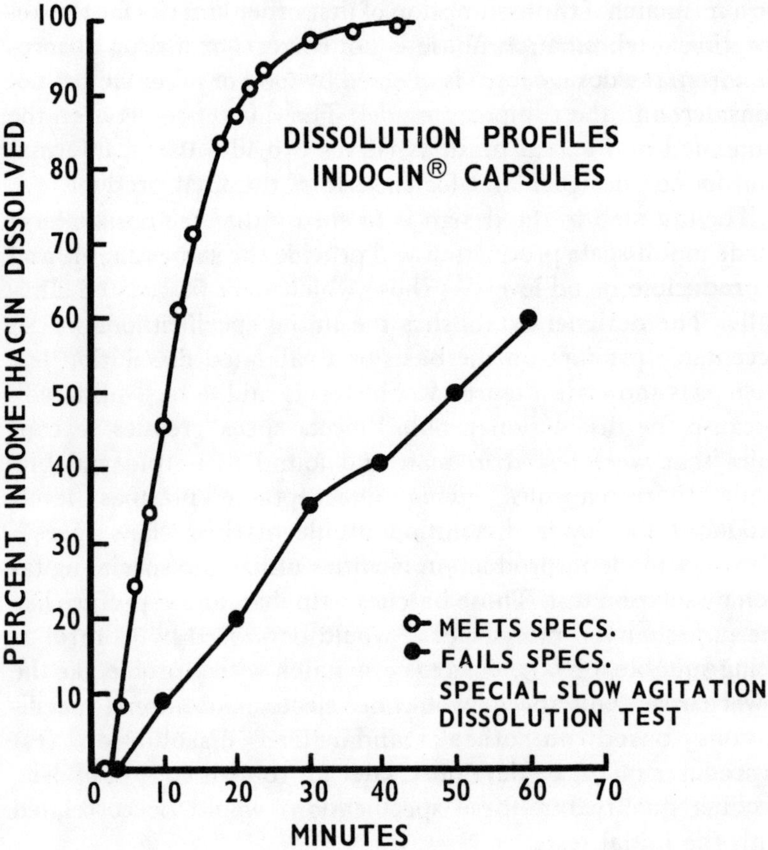

Figure 19.

of providing reliable conception control for a month or more. The drug in the vaginal device is formulated in a physiologically compatible polymer matrix. The rate of drug release is predetermined by proper adjustment of the physical characteristics of the drug-polymer matrix in such a way that the diffusion of the drug molecules from the matrix is slowed to give just the desired rate of drug release. For example, drug-polymer matrix systems have been devised recently in the form of small discs for prolonging the release of drugs to the eye. These thin discs, made of a soft plastic, polymeric substance, are comfortably tolerated when placed under the lower eyelid. Drug effects from such discs may last 10 times longer than from conventional eye drop formulations. The key to the design of these fixed-release rate delivery systems involves detailed knowledge about what constitutes optimal levels of the drug at the target organ site. Also, one must know a great deal about the drug's rate processes (pharmacokinetic parameters) involved in the *absorption, distribution, metabolism* and *excretion* (ADME) in both normal and disease states.

An understanding of the rate processes associated with drug activity has been an essential feature of present-day dosage forms. However, the real legacy of this pharmacokinetic knowledge is not just the list of products found on pharmacy shelves, but the new conceptual approaches being developed for discovering and utilizing the drugs of the future. These approaches owe a great deal to developmental research philosophy in that they are blends of existing and new knowledge applied to solve new problems. Long-established qualitative knowledge from such fields as anatomy and pharmacology are combined with quantitative information from the new field known as pharmacokinetics. For example, it has long been known that venous blood from the area under the tongue (sublingual) is carried by a different pathway than is venous blood from the upper gastrointestinal tract. A number of conventional drugs, such as nitroglycerin, are administered sublingually because this route provides rapid absorption and also because the blood from sublingual tissue goes first to the heart and lungs and then throughout the body before it reaches the liver. This means that a larger portion of the drug reaches the target site before being inactivated by liver enzymes.

Drugs of the future that may be poorly absorbed from the gastrointestinal tract or rapidly inactivated may very well go the sublingual route with special dosage forms that will control drug release at a rate balanced against the known rates affecting distribution and persistence at the site of action. Novel dosage forms, and the concepts underlying their application, will permit pharmaceutical scientists to consider drug candidates that would never be developed with yesterday's technology. Most of today's drugs exert their action over a period of time measured from minutes to hours and in doses ranging from grams, milligrams and occasionally, micrograms. We are just now beginning to study, and consider, drug candidates whose period of effectiveness at the site of action may be measured in seconds and whose effective dosage may range in the order of nanograms (billionths of a gram). One might, with reasonable confidence, predict that in the early 1980s new classes of drugs will be discovered and developed in precisely titrated doses, measured in nanograms, providing benefits in seconds and then leaving the body through predetermined chemical and biological pathways. Their success will depend largely on the ingenuity of the dosage form designer in developing an optimum drug delivery system.

The next generation of "nonconventional" drug delivery systems may very well be represented by a potent but short-acting drug contained in a small device attached to the skin. The drug would be released in direct response to the target organ's need rather than at a fixed rate. Changes in enzyme levels or changes in other physiological compounds may be used as the signal to a feedback mechanism for releasing the drug from an electro-responsive polymer matrix. At the moment, this kind of nonconventional system remains a theoretical possibility, not a practical reality. However, the technological future has a way of happening a lot sooner than most people realize. The designers of today's drug products are not content; they are actively seeking pathways to the drug delivery systems of the 1980s.

Suggested Reading

Schroeter, L.C. *Ingredient X*. New York: Pergamon Press, 1969.

Martin, A.N. *Physical Pharmacy*. Philadelphia, Pa.: Lea & Febiger, 1960.

Martin, E.W. *Remington's Pharmaceutical Sciences*. Easton, Pa.: Mack Publishing Co., 1965.

Wagner, J.C. Biopharmaceutics: Absorption Aspects. *Journal of Pharmaceutical Sciences*, 1961, **50**, 359-387.

Levy, G. Biopharmaceutical Considerations in Dosage Form Design and Evaluation. In J.B. Sprowls (Ed.): *Prescription Pharmacy*. Philadelphia, Pa.: J.B. Lippincott, 1963.

Wagner, J.C. *Biopharmaceutics and Relevant Pharmacokinetics.* Hamilton, Ill.: Drug Intelligence Publications, 1971.

Gibaldi, M. *Introduction to Biopharmaceutics.* Philadelphia, Pa.: Lea & Febiger, 1971.

Notari, R.E. *Biopharmaceutics and Pharmacokinetics: an Introduction*, New York: Marcel Dekker, 1971.

Brodie, B.B. & Haller, W.M. *Bioavailability of Drugs.* New York: S. Karger, 1972.

Clarke, F.H., Editor. *How Modern Medicines Are Discovered.* Mount Kisco, N.Y.: Futura Publishing Company, 1973.

CHAPTER 7

Clinical Evaluation and Safety

Murray Weiner

The administration of a new chemical entity to human subjects is rarely undertaken as a pure basic research effort without aiming at the development of a useful product. In fact, some might question the morality of studies in man conducted solely for the sake of scientific information if a potential for developing a new useful product or improving an existing one were not the major motivation. In this respect, industry-sponsored clinical drug studies are less subject to criticism than some of the activities of more basic research-oriented academic clinical scientists. The concept of "benefit/risk" ratio permeates the thinking and experience of the pharmaceutical research scientist in the very conduct of clinical studies as well as the evaluation of new therapeutic agents and devices.

In spite of, or perhaps because of the goal-oriented nature of industry-sponsored clinical evaluations of drugs, those who undertake this phase of drug development must know what to expect and also be prepared for the unexpected.

By and large, fundamental mechanisms of physiologic action are quite similar from one species to another, and pharmacologic effects are likely to be at least qualitatively similar, so that clinical studies must be based on animal observations for both the safety of the human volunteers and the probability of clinical research success. However, there are abundant examples of marked species differences that were not predicted in advance or even explainable with hindsight. Thus, penicillin kills guinea pigs in doses per unit weight much smaller than ordinary clinical doses in man. Yet the guinea pig is uniquely useful for other evaluations, such as studies involving Vitamin C. With some drugs (anesthetics and antimicrobial agents) a comparison of blood levels achieved rather than doses administered does away with much of the species differences and improves predictability in man. But in many instances, even after species differences in rate and route of metabolism and protein-drug complexing are taken into account, considerable species variations remain. Invariably, diseased man must be the ultimate test animal.

There are innumerable examples in which the interplay of keen clinical observation and evaluation with good laboratory science has led to practical achievements that neither could have accomplished alone. The concept of histamine release as a prime component in explaining many allergic reactions was the consequence of brilliant experimental observations in animal studies correlated with chemistry. It was inevitable that antihistamine drugs would be sought and found. Among the compounds made and clinically tested as antihistamines was chlorpromazine, the granddaddy of a whole family of tranquilizers that has reduced by half the census of our mental institutions. Just as the tranquilizing properties of chlorpromazine were discovered during its clinical evaluation as an antihistaminic, so the antidepressant properties imipramine were discovered during its clinical evaluation as a possible tranquilizer. The original synthesis of imipramine had also been undertaken with antihistaminic activity in mind, and its emergence as a tranquilizer was stimulated by reports of the observations with chlorpromazine.

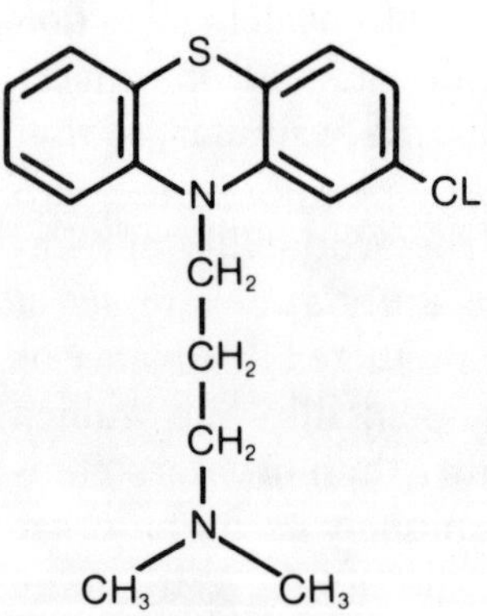

Chlorpromazine

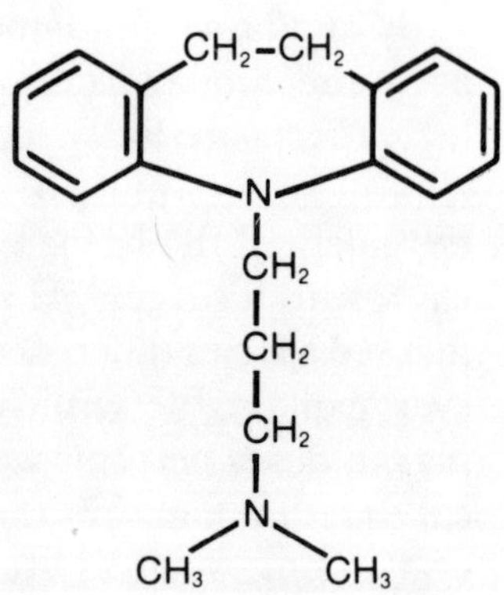

Imipramine

The sequence of events in the development of imipramine, as seen by a U.S. research team from Geigy Pharmaceuticals, is particularly interesting. Before it was aware of the existence of the chemical substance now known as imipramine, this group had investigated its isopropyl analogue as a tranquilizer and found it to be active but weak. They found no potential clinical advantage over chlorpromazine, and decided not to pursue it further. They were, therefore, surprised at the suggestion from their associates in Europe that the n-propyl analog (imipramine) be studied, and for a quite different indication, i.e., depression, for which there was no supporting animal data. In fact, there was no animal model known that reflected antidepressant activity. The suggestion was based purely on clues derived from astute clinical observations in one European clinic. Eventually, the validity of these observations was fully confirmed in U.S. clinics.

The clinical observation that was the key to this development could easily have been overlooked if the investigator had neglected to think beyond the limited initial objective of his study. He and the cooperating research organization started with the straightforward objective of determining how this experimental compound compared to chlorpromazine as an antipsychotic in schizophrenic patients. A modest clinical experience was enough to suggest that imipramine was of little value as an antipsychotic. The cardinal signs of psychosis, such as hallucinations, did not improve. In fact, in a few quiet, withdrawn schizophrenics, the drug seemed to convert them to a more troublesome behavior pattern in that they became less withdrawn and were now reacting to their hallucinations. The clinical investigator, R. Kuhn of Musterling, Switzerland, could have categorized such patients as "worse" and given up the drug. Instead, he made the observation that it effectively influenced the depressed component of the symptom complex in these patients. He therefore decided to reevaluate the agent in patients who were primarily depressed and not necessarily schizophrenic. The results convinced him that the drug was clearly antidepressant.

Imipramine now became a tool for searching out an animal

model for human depression. At first, attention was drawn to the then novel and unexpected observation that the new tranquilizer, reserpine, was capable of inducing rather classical clinical depressions. Since reserpine was also characterized by a pattern of readily observable and reproducible pharmacologic responses in animals, such as eyelid closing (ptosis) and hypothermia, it was reasoned that the capacity of a drug to prevent or reverse these reserpine-induced effects in animals might reflect a clinical antidepressant potential. The approach was even more intriguing because the pharmacologic effects of reserpine had been postulated to be associated with distinct effects of reserpine on several endogenous neurohumoral conductors. The observation that pretreatment with imipramine potentiated the effects of norepinephrine, resulted in a series of laboratory models that made possible the screening and eventual clinical study and marketing of a whole series of effective new antidepressants, even though, to this day, no one model is a sure-fire predictor of clinical antidepressant action. Drug development was an invaluable aid in the discovery of new basic concepts about emotional diseases and normal neurochemistry, and these new concepts in turn were critically important to the development of new drugs.

Until these discoveries there was essentially no basis for designing and evaluating animal models to help select compounds that might be efficacious in this clinical area. It was only a scant few years before the discovery of imipramine that depression as a psychiatric entity came to be distinguished from the nondiagnosis of "nervous breakdown." The discovery in the late 1950s that depression can be lifted by agents that inhibit the enzyme monoamine oxidase was a major stimulant to the clinical recognition of depression as a common disease. Defining and understanding a particular disease pattern is often stimulated by the development of a new therapeutic approach, which brings new importance to the need for better diagnostic methodology and a clearer image of the nature of the underlying disturbance.

The preoccupation with monoamine oxidase inhibition as the presumed mechanism of antidepressant action in the late 1950s was in one sense a handicap to the development of imipramine as an antidepressant since imipramine is not an MAO inhibitor. Nevertheless, the weight of controlled clinical studies confirmed its clinical antidepressant activity, and improved theories concerning its mechanism of action led not only to useful animal models but to a better understanding of brain function in general. Major studies in neurochemistry were the outgrowth of these drug discoveries. The concept that imipramine was not an MAO inhibitor was not readily accepted. When animal and *in vitro* studies showed no such inhibition, there was still the hypothesis that in man the drug might be converted metabolically to an MAO inhibitor. Eventually, definitive tests proved this false. At that point, some scientists who had embraced the earlier theories relating MAO inhibition and antidepressant action began to question the clinical observation that imipramine is antidepressant. By then the results of observations under well-controlled conditions were undeniable, and it became necessary to develop alternative biochemical hypotheses and pharmacologic models. The clinical empirical fact was now the axiom on which theories had to be based, and not the other way around.

The schematic diagram (Fig. 1) of the sites of action of drugs affecting norepinephrine at adrenergic nerve terminals evolved largely as a direct consequence of investigations of the mechanism of action of clinically useful, industry-developed agents. This new-found knowledge in turn made it possible to design new laboratory models that industry has employed to help develop a new series of significant candidates for clinical investigation. Hundreds of thousands of people benefit each year from the fruits of this sequence of events.

Another example of interplay between industry-sponsored research, sound, basic science and skillful, imaginative clinical studies, has been the development of new agents for the treatment of chronic gout. Three useful agents in this area (probenecid, sulfinpyrazone and allopurinol) came about from studies with pharmaceutical industry cooperation and support aimed at other purposes.

Probenecid was originally of interest in inhibiting the excretion of penicillin by the kidney and was found in the clinic to

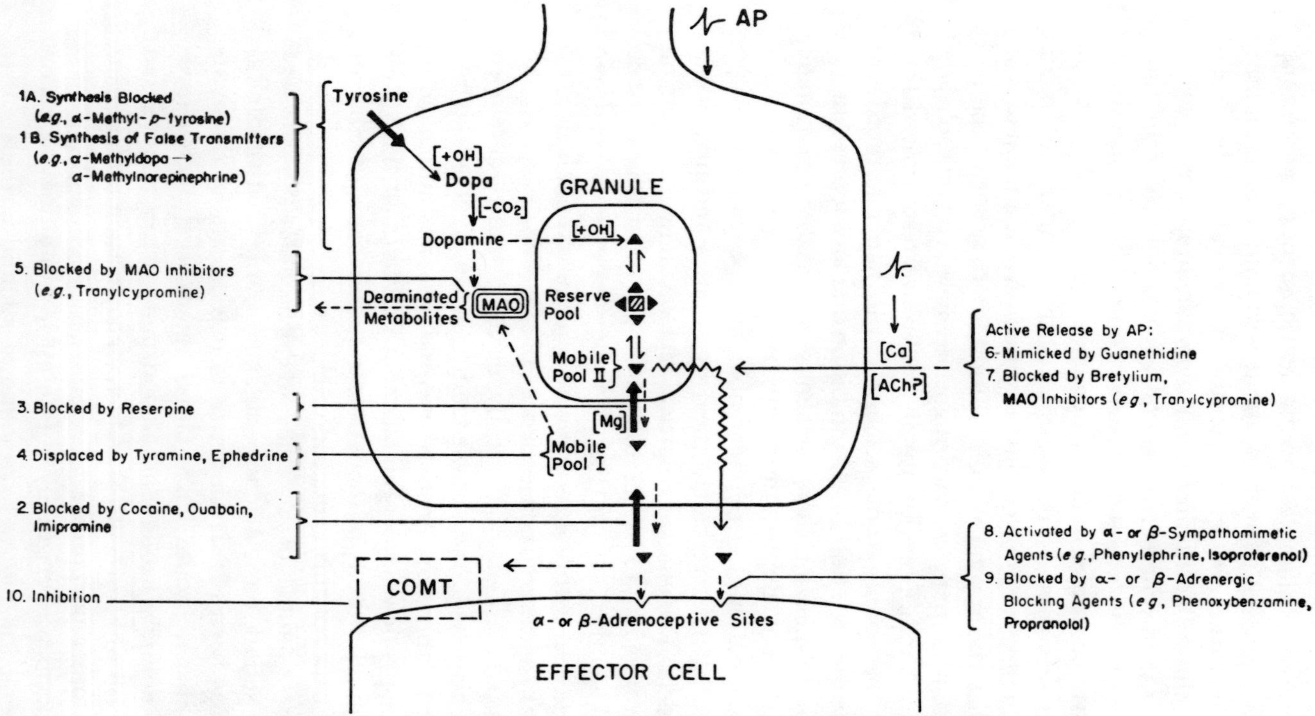

Figure 1. Drug mechanisms at adrenergic nerve terminals. From *The Pharmacological Basis of Therapeutics*, Fourth Edition, by Louis S. Goodman and Alfred Gilman, The Macmillan Company with permission of the publisher.

enhance the renal excretion of uric acid, a useful phenomenon in controlling chronic tophaceous gout. Years later, a similar though much weaker effect was observed with a totally different drug, the anti-inflammatory agent, phenylbutazone. This compound was clearly effective in suppressing the acute inflammatory component of arthritic diseases, including gout. While it was recognized that the relief of acute gouty arthritis was not correlated with the successful treatment of high serum uric acid levels associated with chronic tophaceous gout, it was not surprising that the early investigators of phenylbutazone studied the effect of the drug on serum and urinary urates. For some time the reports were conflicting. It was later found that high doses of phenylbutazone were uricosuric, i.e., enhanced the urinary excretion of uric acid, even though lower doses, which were quite adequate to control acute gout, were not uricosuric. Eventually, each of two metabolites of phenylbutazone were identified, synthesized and clinically tested. Metabolite II proved to be strongly uricosuric when injected. It was not anti-inflammatory, was poorly absorbed orally, and had a short plasma half-life. Clinical investigation of absorption and uricosuria in a few volunteers with a series of analogues synthesized by a pharmaceutical research laboratory established a correlation of a physical-chemical characteristic (pKa) of the drug with its action on uric acid excretion. This basic fact made it possible to design deliberately a new series of drugs for more detailed study. As a result, one prime candidate successfully studied in man yielded a urinary metabolite that was isolated, identified and, eventually, shown to be even more active and specific than the parent compound. This drug, sulfinpyrazone, was developed into a useful uricosuric agent that became commercially available.

Since patients with chronic tophaceous gout tend to have a high incidence of thrombovascular disease and abnormally sticky platelets, the response of these factors to sulfinpyrazone came under scrutiny. The evidence showed that this drug inhibits platelet aggregation and normalizes the abnormally short half-life of platelets in patients with atherosclerosis and related thrombovascular disease. The seeds of further development of sulfinpyrazone or other agents with similar action have been planted, and may some day be harvested as new rewards in health and longevity in yet another area.

Allopurinol provides a wholly different approach to the treatment of chronic gout, which involves an inhibition of the synthesis of uric acid, rather than its enhanced excretion. This drug was originally prepared to prolong the action of an anticancer drug. It is now used for both purposes. It is an example of a drug designed for a specific purpose based on fundamental knowledge. An enzyme responsible for the destruction of a useful cancer agent was identified and studied. Compounds were deliberately designed and tested to inhibit this enzyme. The plan and its development was ingenious and successful. It became apparent that this same enzyme inhibitor blocked the synthesis of uric acid. It is particularly useful where the high urate concentrations in the urine predispose to the formation of urate kidney stones. The other drugs (probenecid and sulfinpyrazone) lower blood uric acid by washing it out through the kidneys, thus exposing the kidney to large amounts of urates. Allopurinol blocks the synthesis of urates and therefore markedly reduces the amount going through the kidney.

From these examples it is clear that new drugs must be evaluated in man, and at times their utility cannot be fully appreciated or even discovered until they are studied in man. Any activity undertaken for the first time in man carries with it the risk of the unknown. Scientists responsible for the first study of a new chemical entity in humans are acutely aware of the need to be guided by the experience that has led to the judgment that a particular compound deserves to be studied in man. The coordinated efforts of several disciplines precedes a decision that the nature of a chemical entity and its known action in animals warrants its testing in humans for a particular purpose. Such first-time activity is developed with a multitude of safeguards, not the least of which is an awareness of the probable nature of the potential risks involved.

While our system of new drug discovery is founded primarily on observations in animals and in biochemical laboratories, the value of this foundation can be eroded if adequate experience and wisdom are not applied to the design of clinical studies. The major resource of existing scientific talent with training, experience and interest in this vital area is to be found in the pharmaceutical industry. The industry has built a remarkable series of competing interdisciplinary teams, each of which has the facilities, skill and centralized goal-oriented scientific direction and motivation for fruitful pharmaceutical development. Today's clinical research scientists in industry are respected members of their professional communities and are often welcomed as teachers, consultants and advisers in medical schools and learned societies.

Whenever a talented scientist who can contribute to drug development appears on the academic scene, he is bound to be sought out by his industrial colleagues for whatever contribution he can make as investigator, co-worker, or consultant. After all, both have a primary goal—the development of safer and more efficacious drugs. This fact again reflects the vitality of the industrial research effort. Its continuing success is of prime interest to the medical community and the public welfare. It is clearly in the public interest not to exclude from government advisory groups every scientist who ever worked for, or consulted with, the pharmaceutical industry. Such exclusion would deny these agencies access to those scientists who are in the best position to make informed judgments. The "revolving door" for academia, industry and government need not be a conflict of interest but rather a mechanism for fuller understanding in the cause of better drug therapy. A well-trained, morally-sound scientist does not become a knave on entering industry or a fool on entering government service. Scientific truth and conscience are not the exclusive possessions of any one sigment of society. They are pursued by all scientists worthy of the name.

Government regulating agencies by law assume responsibility for public safety in overseeing the clinical evaluation of drugs. The FDA has a vital role in protecting the patient and the legitimate investigator against the excesses of the occasional individual who may make his way into a position for which he is not suited. Similarly, the press discharges its responsibilities well when it exposes such situations. However, legitimate scientific differences of opinion do arise among competent scientists. The public interest is poorly served when the FDA uses its legal power, or the press editorializes, to compel the entire medical community to adhere to one viewpoint, even though expert opinion may be still divided.

Few things are as insidiously dangerous to the public welfare as to have the moral and scientific judgment of competent clinical investigators influenced by concern for how their statements might be misrepresented or distorted. Judgments in the design and conduct of clinical evaluations obviously should be made objectively, with concern for what is best for the patient and not out of fear of being misunderstood or deliberately distorted. Defensive research, designed to avoid criticism at the expense of logic and good science, poorly serves the public interest. The nature and timing of literally scores of biochemical and clinical pharmacologic parameters need to be considered individually for each drug and each situation. It would indeed simplify the life of industry scientists if some omniscient body could write out infallible rules for them to follow. But no such body exists, and no such rules are possible. The safety of patients and the fruitfulness of research would be reduced to the vanishing point if they were to work strictly "by the numbers." Both drugs and people are too different from each other for any one set of rules to be blindly applicable. All things considered, the man on the spot with the greatest knowledge is in the best position to protect patients and extract the most information from the study.

Thus, while a wide variety of detailed "guidelines" have been developed recently under which new drugs are to be investigated in man, no guideline can ever substitute for the skilled scientific and moral competence of the people involved. At the present time, one real handicap to developing significant clinical information, at the least risk to patient volunteers, is the frequent dilution of important observations with a multitude of routinized demands for making and recording observations of limited value to a given stage in the investigative procedure. While the objective is enhanced safety by trying to cover every conceivable eventuality, there is a breaking point at which patient welfare can be smothered by "safeguards." The public needs to understand, and the opinion molders have a responsibility to help them understand, a pattern of relevant facts. Much is written about harmful reactions to drugs, but not enough is said about the thousands now in their graves who might be alive if the development of promising drug leads had not been hampered to the degree where their potential values were never realized. We are reaching a point where the blind and unqualified demand for safety may become the most dangerous threat to extending the quality and duration of life by advances in medical science.

Another public threat to the ongoing search for new medicines is the critic with an ax to grind or a reputation to make. The true hero in facing up to this problem is the scientist who would rather be right than popular. The worst disservice a clinical investigator can do for himself and for the industrial sponsor is to report encouragingly on a drug which is, in fact, not safe and effective. It is advantageous to all to find reasons why a presumably promising compound should be cast aside. The assumption that industrially sponsored research is motivated to report positive findings where they do not exist simply is not true. It makes no sense from either a scientific or company standpoint.

It is beyond the purpose of this chapter to detail the extensive recent activity in this country and worldwide that spells out the moral considerations and precautions that the trained clinical investigator must consider. Reams of specific guidelines have been developed to make clinical research both safer and more effective. Their mere development has been a worthwhile education, and the teaching of clinical investigation in departments or subdepartments of pharmacology has now become the rule rather than the exception. But we must never succumb to the notion that formalized guidelines, no matter how brilliant, can substitute for professional and moral competence that every commercial sponsor seeks in his own enlightened self-interest. The premise that he is motivated to hide negative information is neither true nor constructive. The cost of drug development is so great that he cannot afford the luxury of misguiding information. In the development of new drugs, the goals of the pharmaceutical industry have to run parallel to the interests of the public.

There is a danger of eroding the creativity of the competitive pharmaceutical development process by a steady shift to first accommodate regulatory requirements, even if inappropriate and unrealistic, rather than sound scientific and moral reasoning. This trend is often accompanied by a nagging legal-type adversary position between scientists in industry and in regulatory agencies. What is needed is more of a clinical scientist-type relationship that seeks truth by objective evaluation without the black-white, yes-no quality generated by the adversary approach. Industry and the agencies can serve the public best if there is recognition of the mutuality of their interests along the difficult path to fruitful clinical evaluation of new drugs, devices and diagnostic products.

Suggested Reading

Kline, D.F. and Davis, J.M., "Diagnosis and Drug Treatment of Psychiatric Disorders", Williams and Wilkins, Baltimore, 1969.

Lasagna, L., Editor, "Clinical Pharmacology", Section VI, *International Encyclopedia of Pharmacology and Therapeutics*, Permagon Press, Oxford, 1966.

McMahon, F.G., Editor, "Future Trends in Therapeutics" in Tulane Medical Symposium, November 2-5, 1975, "Principles and Techniques of Human Research and Therapeutics" (in press).

Tedeschi, D.H. and Tedischi, R.E., Editors, "Importance of Fundamental Principles in Drug Evaluation", Raven Press, N.Y., 1968.

Weiner, M. & Piliero, S. Non-Steroid Anti-inflammatory Agents. *Annual Review of Pharmacology*, 1970, **10**, 171-197.

Weiner, M. An Immunologic Approach to the Understanding of the Physiologic Disposition of Drugs. *Life Sciences*, 1973, **13**, 1473-1481.

USP Conference on Bioavailability. *Pharmacology*, Vol. 8, 1972.

Weiner, M. The Objective and Timing of Drug Disposition Studies. *Drug Metabolism Reviews*, 1975, **4**, 229-239.

CHAPTER 8

Developmental Research and the Future

George deStevens

The previous chapters have outlined with specific examples how a substance, once identified as having potential medical utility, is developed further so that it can become part of the doctor's armamentarium. This has been a process long familiar to the pharmaceutical industry, and I would like to give as a prelude to the future, two examples of how developmental research is so vital in drug research.

One classic example is the development of cortisone. The story goes back to 1948 when P.S. Hench and E.C. Kendall of the Mayo Clinic had requested from Merck scientists a small quantity of cortisone for the treatment of rheumatoid arthritis. The synthesis of cortisone had already been carried out in the research laboratory, so the next task was to obtain larger quantities in the development laboratory. Consequently, Dr. Max Tishler, at that time Director of Developmental Research for Merck & Co., and his colleagues worked most painstakingly to obtain a few grams of cortisone via a 52-step synthesis. The subsequent clinical work by Hench and Kendall was revolutionary, since it showed that a synthetic substance could be useful in the treatment of arthritis.

The first batch of cortisone for clinical trials cost approximately \$200 per gram. As soon as the therapeutic value of cortisone was determined, Merck scientists in developmental research turned to all of their know-how in chemistry to develop a synthesis which would give higher yields of cortisone at a much lower price. Within one year, these researchers had developed a 20-step synthesis that was applicable on a pilot plant scale and reduced the cost to \$10 per gram. This type of research not only made cortisone available in larger quantities for clinical studies but was also largely responsible for the development of new and improved corticosteroids for the treatment of rheumatoid arthritis. This developmental research, which established the basis for corticosteroid synthesis, was carried out in the late 1940s under what one could consider a *crash* program. It demonstrated the enormous capacity of the pharmaceutical industry to transform discoveries in clinical research into practical and useful therapy. (Incidentally, the present selling price for bulk cortisone is $.38 per gram.)

Another approach to developmental research took a much longer time. During the mid 1950s, the research management of CIBA Limited decided that an important area of therapeutics within the next 20 years would be hormonal peptides. These hormones are not available in quantities which could be extracted from animal organs for clinical trials. Thus it became clear that any commercial and medical success in this area would have to be controlled by the degree of sophistication that could be established in peptide synthesis in the laboratory. With this challenge in mind, in the mid 1950s, the CIBA organization committed itself to a basic chemical research program in peptide synthesis. This meant that the management, on the recommendation from research, was prepared to invest heavily in order to develop the know-how in the laboratory that could be translated in a practical way to the manufacturing level. It was also necessary for both research and management to be convinced that it would be at least a decade before there could be a so-called return on the investment. Within a period of seven years, the peptide chemists in the CIBA organization developed the expertise for the synthesis of ACTH, a 39-amino acid polypeptide hormone, and also a 24-amino acid peptide, and a variety of other related ACTH-like molecules. Polypeptides (or proteins) are long chain molecules formed by the head to tail combination of sequences of the 20 naturally occurring amino acids. Links in the chain are formed when the amino group of each amino acid joins with the acid group of its predecessor to form an amide or peptide bond.

ACTH (corticotrophin) is one of the hormones which causes the production of adrenal cortical hormones, and thus offers an alternative to the corticosteroids, such as cortisone, for therapy of inflammatory diseases. The product that resulted from this developmental research was eventually introduced in a number of countries in Europe under the trademark Synacthen®, which is a peptide made up of 24-amino acids and has all of the biological activity of natural ACTH. The reception of this drug by the medical community in the countries in which it was introduced was gratifying. For the first time the doctor could use this pure substance in exactly measured milligram quantity rather than in the traditional international units ascribed to the ACTH extract.

Although it is too early to determine the full commercial impact that Synacthen® will have, CIBA-GEIGY felt that this research had led to a valuable product. As to scientific value, this research has made possible the recent announcement by CIBA-GEIGY scientists in Switzerland that, after 15 years of research, they have finally proven the exact chemical structure of *human* ACTH. W. Rittel and his group substantiated this structure proof by total synthesis.

A further consequence of CIBA-GEIGY peptide research is the present pioneering work being done with the hormone calcitonin. Here again, this most important substance had been practically unavailable. Scientists have proven the structure of human calcitonin and now have made available, though in very small quantities, the pure synthetic hormone to a number of clinical investigators throughout the world. Calcitonin is the only substance which is effective in the treatment of Paget's disease, an ailment in which the bone is rendered structurally weak and prone to deformity and fracture. It is estimated that there are only about 2000 people suffering from this dreadful disease in the United States. Nevertheless, CIBA-GEIGY feels a responsibility toward their well-being. In addition, further developmental research in modification and new analogs of the calcitonin peptide molecule may lead to further basic knowledge. This commitment to time and material has only been justified because the chemists in developmental research were able to develop new synthetic procedures that make the synthesis of these peptides practical for clinical evaluation and further medical research.

The examples just described reflect what this industry is

capable of doing once the medical need is established. Now to a few thoughts on the role of developmental research tomorrow.

In the future, new drugs will be used to cure rather than merely control complex disease states. This means that the basic developmental knowledge must come primarily from biology. For the past decade there has been an increasing emphasis on molecular biology, and within the past few years its contribution to basic biological determination of *controlled* life function has led us to a closer understanding of the *uncontrolled* life process commonly known as cancer. The key material is DNA, the master chemical of life. The DNA of the genes, through a sister substance called RNA, controls the structure and function of the cell, that is, the synthesis of proteins.

The techniques of molecular biology make it possible to find out, in great detail, what is happening even at that complex submicroscopic level. Each gene prescribes a unique messenger RNA. Strands of it correspond to strands of the genetic DNA that spell out the sequence of chemical bases in the RNA. Scientists can fish various messenger RNA's out of the soup-like interior of the cell by employing the DNA that produced them as a sort of magnet. That enables the researchers to determine whether a particular protein was produced by the cell's DNA or by the DNA stemming from a virus. In this way, the association of a specific virus with a specific cancer can be established.

Scientists have made some imaginative attempts to explain why certain viruses produce cancer—and why viruses exist at all. One such hypothesis was proposed by Howard M. Temin of the University of Wisconsin. In his concept of what goes on, there exists in the genetic system of cells, a special region of DNA whose job it is to transmit genetic information from cell to cell, mainly during the embryonic development of an organism. Temin calls this gene system a "protovirus." In normal development of the organism, the DNA protovirus of a cell would act as a template for the synthesis of an RNA copy of itself. The RNA copy—not enclosed in a coat like full grown virus, but "packaged" with the enzyme needed to translate it back into DNA—would then enter another cell and there act as a template for a synthesis of a DNA copy of itself. This DNA copy would then become incorporated into the genome of the second cell, imparting to it the message carried in the transferred pieces of nucleic acid.

If Dr. Temin is right, the protovirus is genetic dynamite, for errors may occur during this information transfer that may be beneficial, indifferent or damaging. According to the Temin hypothesis, such errors can lead to cancer or, at least certain types of cancer; when the protovirus is altered, its genes can start altering the cell's responses to growth controls. This hypothesis may explain why such diverse factors as viruses, chemicals and radiation all can produce cancer cells of similar types. Viral genes that penetrate a cell may have a special chemical affinity for the protovirus region and the damage may be concentrated there. Chemicals and radiation may similarly affect the protovirus. Some cells are killed while others survive with the protovirus genes damaged. In Temin's hypothesis, then, cancer is not a purpose of the protovirus but an accidental effect. He views disease causing viruses as abnormal variants of the protovirus particles.

Somewhat similar to that of Temin's is an older hypothesis put forward by Robert J. Huebner and George Todaro of the National Cancer Institute. They suggest that the cells of all or most vertebrates may carry as part of their evolutionary inheritance a normally suppressed set of "oncogenes," or cancer genes. In this hypothesis, these can produce cancer when subjected to abnormal stimulation.

A most exciting sequel to the hypothesis suggested by Huebner and Todaro, is the identification of a new virus, known as RD-114. This virus, isolated by Robert M. McAllister and Murray B. Gardner of the University of Southern California School of Medicine is believed to be a prehistoric virus, and new evidence suggests that it may be the cause of most human cancers. This virus, which is imperceptibly blended with the human DNA and believed to be 500,000 years old, is claimed to be part of our inherited genes.

To sum up, this virus in not infectious but appears to pass on cancer by viral heredity. I don't think it is too optimistic to say that this work will lead to a cure for several forms of cancer within the next 15 years.

An essential fallout from this research will be that viruses also will be used to pass on to the cell the information that will prevent diseases due to metabolic derangement. Two examples of this would be the cure of diabetes and rheumatoid arthritis.

Finally, I would like to briefly outline future developments in the area of drug design. I recall a story in which Sir Alexander Fleming was the central figure. One day in 1922 he was suffering from a cold. This is not unusual in London, but Sir Alexander was a most unusual man and he took advantage of the cold in a characteristic way. He let a few drops of his nasal mucous fall on a culture of bacteria he was working with and then put the plate to one side to see what would happen. Imagine his excitement when he discovered sometime later that the bacteria near the mucous had dissolved away.

For a while he thought his ambition of finding a universal antibiotic had been realized. In a burst of activity he quickly established that the antibacterial action of the mucous was due to the presence in it of an enzyme. He called this substance lysozyme because of its capacity to lyse or dissolve the bacterial cells.

Lysozyme was soon discovered in many tissues and secretions of the human body, in plants and, most plentifully of all, in the white of egg. Unfortunately Fleming found that it is not effective against the most harmful bacteria. He had to wait seven years before a strangely similar experiment revealed the existence of a genuinely effective antibiotic: penicillin.

Nevertheless, Fleming's lysozyme has proved a more valuable discovery than he could have expected when its properties were first established. As it has turned out, lysozyme is the first enzyme whose three-dimensional structure has been determined and whose properties are understood in atomic detail. Among these properties is the way in which the enzyme combines with the substance on which it acts—a complex sugar in the wall of the bacterial cell.

Like all enzymes, lysozyme is a protein. Its chemical makeup had been established by Pierre Jolles and his colleagues at the University of Paris and by Robert E. Canfield of the Columbia University College of Physicians and Surgeons. They have found that each molecule of lysozyme obtained from egg white consists of a single polypeptide chain of 129 amino acids.

The mucopolysaccharide structure found in the walls of certain bacterial cells is the substance broken by the lysozyme molecule. It is a polymer composed of alternating residues of two kinds of amino-sugar: N-acetyl-glucosamine and N-acetyl-muramic acid. For the lysozyme to be effective in cleaving the polysaccharide substrate, the latter must interact with the enzyme in a most precise manner. This is because the lysozyme molecule is folded together in such a way that a valley or cleft exists that has a shape exactly complementary to the polysaccharide. The valley formation occurs not by chance but rather as the result of the many binding forces within the total molecule. Thus the evolutionary process has been such that the protein molecule, lysozyme, is so developed that the cleft is prepared to receive only the polysaccharide of cell walls in a perfect fit. Once that has taken place, the particular responsible molecule groupings are properly oriented to cause the catalytic cleavage of the polysaccharide chain.

This type of action at the cellular level occurs in all living systems. That is to say, the millions of reactions responsible for proper cell functioning are generally those whereby a substance which is to be transformed must fit into a particular receptor site of a protein molecule, be it enzyme or otherwise.

Drugs also act in the same way. A drug, once absorbed into the blood stream, is carried to its site of action. The drug receptor site may be an enzyme, as it may be the locus where a body hormone produces its normal effect. The task of drug research in the future will be to understand the intimate structural features of the drug receptors and thus to ascertain the exact site where the drug substance is effecting its action. With this information in hand, the chemist in the laboratory can design the drug molecule in a precise way, so that it will fit the receptor site according to specification.

Such work is currently being carried out in many academic

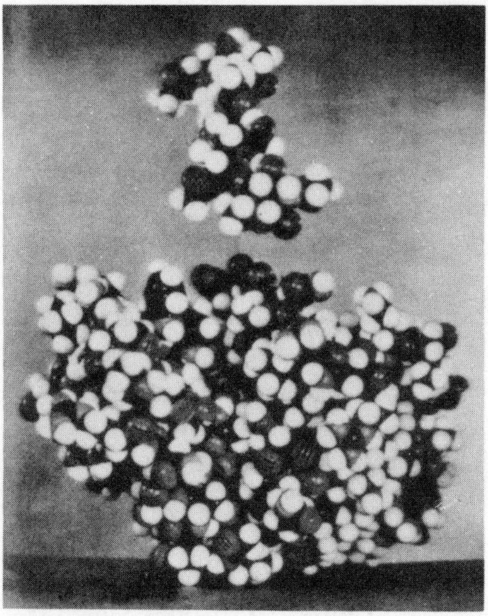

Figure 1. Enzyme model. Substrate poised above the active site of lysozyme.

Figure 2. Enzyme-substrate interaction. The substrate of Figure 1 fills the active site of the enzyme.

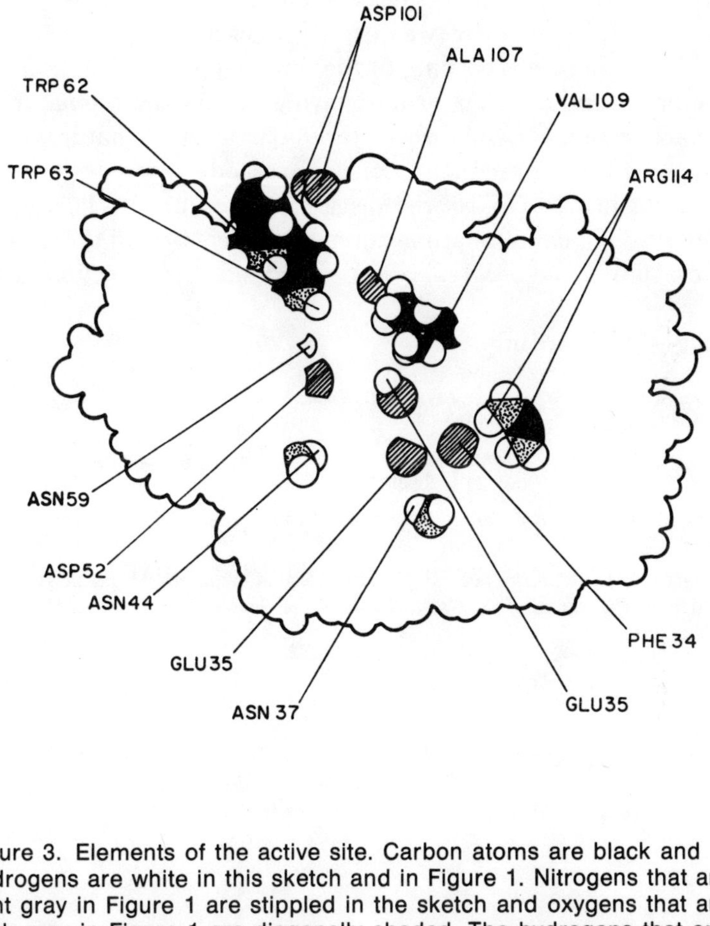

Figure 3. Elements of the active site. Carbon atoms are black and hydrogens are white in this sketch and in Figure 1. Nitrogens that are light gray in Figure 1 are stippled in the sketch and oxygens that are dark gray in Figure 1 are diagonally shaded. The hydrogens that are participate in hydrogen bonds with substrate are specified by the points of termination of identifying lines. Figures 1, 2 and 3 have been adapted from R.A. Harte and J.N. Rupley, "Three-dimensional Pictures of Molecular Models" in *The Journal of Biological Chemistry*, Vol. 243, pp. 1663-1669 (1968). The original article contains magnificent three-dimensional pictures in full color.

and industrial research laboratories, and it has far-reaching consequences. As a result of the precise design of drug molecules, the drug will be much more efficient in providing its beneficial effects on a target cell. The toxicity that plagues so many drugs today will be greatly reduced. The amount of drug to be given a patient will no longer be a matter of tradition or conjecture, such as one tablet two or three times a day, but will be exactly measurable according to the disease entity being treated.

Finally, and most important, having understood the nature of the active site, we will also better understand the nature of the disease entity. Consequently, the precisely designed drug will be used not just to control the disease but to cure it. This is the present-day goal of pharmaceutical research, and in the next 25 years significant progress will be made in this direction.

Suggested Reading

Dickerson, R.E. & Geis, I. *The Structure and Action of Proteins.* Reading, Mass.: W.A. Benjamin, Inc., 1969.

Phillips, D.C. The Three-dimensional Structure of an Enzyme Molecule, *Scientific American*, 1966, **215**, 78-90.

Rang, H.P. *Drug Receptors—A Symposium.* Baltimore, Md.: University Park Press, 1973.

Todaro, G.J. & Huebner, R.J. The Viral Oncogene Hypothesis: New Evidence. *Proceedings of the National Academy of Science, U.S.A.* 1972, **69**, 1009-1015.

THE AUTHORS

Dr. Frank H. Clarke was born in Newcastle, New Brunswick, Canada. He received his B.Sc. and M.Sc. in Organic Chemistry from the University of New Brunswick in 1949 and 1950 respectively. In 1954 he received his Ph.D. in Organic Chemistry from Harvard University. After two years of post-doctorate research at Columbia University he joined the Medicinal Chemistry Department of Schering Corporation where he became Senior Research Chemist in 1959. He joined Geigy Chemical Corporation in 1962 where he was appointed Director of Medicinal Chemistry in 1967. In this position he was responsible for the Departments of Organic Chemistry, Biochemistry and Molecular Biology and for the Kilo Laboratory. He is presently Deputy Director of the Chemistry Division in the Research Department of the Pharmaceuticals Division, CIBA-GEIGY Corporation.

Dr. Clarke's publications are mainly in the areas of the synthesis of natural products and of potential medicinal agents. His interests include the use of drug metabolism in the design of new medicinal agents and the study of enzymes as models for pharmacological receptor sites. He has developed a new color coded skeletal-space filling molecular model system especially suited for building models of enzymes. He edited "How Modern Medicines are Discovered" and is Editor-in-Chief of "Annual Reports in Medicinal Chemistry".

Dr. George deStevens was born in Tarrytown, New York. He received his Ph.D. in organic chemistry from Fordham University in 1953. The title of his thesis was "The Chemistry of Lignin and Biochemical Implications of this substance in Woody Tissues". This work was carried out under the mentorship of Professor F. F. Nord. Dr. deStevens joined CIBA Pharmaceutical Company in 1955 as a senior chemist and was promoted to Director of Medicinal Chemistry in 1961. He became Director of Chemical Research in 1966 and in June

1967 was promoted to Vice President and Director of Research. He is presently Executive Vice President and Director of Research of the Pharmaceuticals Division, CIBA-GEIGY Corporation.

Dr. deStevens has authored and co-authored approximately 100 scientific papers and holds 65 patents in the area of medicinal chemistry. He is the author of a book on diuretics and editor of another volume on analgetics. Dr. deStevens is also editor of a series of monographs on Medicinal Chemistry.

Dr. Irwin J. Pachter, a native of New York City, received his professional education in organic chemistry at UCLA (B.S.), University of New Mexico (M.S.) and University of Southern California (Ph.D.) followed by post-doctoral studies at Illinois and Harvard Universities. He worked on natural products and diuretics for seven years as a chemist at Smith Kline Corporation, for seven years at Endo Laboratores (Div. of DuPont Co.) where he was Director of Research and is now in his seventh year at Bristol Laboratories (Div. of Bristol-Myers Co.) where he is Vice President for Research and Development. He has made notable contributions to the discovery and development of drugs for relief of pain, central nervous system disorders and infectious diseases.

Dr. Ira Ringler was born in New York City and received his academic training at Ohio State University (B.S.—Agricultural Chemistry) and Cornell University (M.N.S.—Nutrition and Ph.D.—Biochemistry). After two years of post-doctorate research at Harvard University, he joined Lederle Laboratories in 1957 as a senior scientist. He served in several administrative positions in the R&D organization, and was named Director of Research in 1969. Dr. Ringler moved to Abbott Laboratories in 1975 and presently is Vice-President for Corporate Research and Experimental Therapy.

Dr. Louis C. Schroeter was born in St. Louis, Missouri. He received his B.S. (Pharm.) (1952) and M.S. (1956) from the St.

Louis College of Pharmacy, his Ph.D. (Pharm.) from the University of Wisconsin (1959). He has served on the faculty at St. Louis College of Pharmacy. He joined the Upjohn Company in 1959 and has served in a number of technical supervisory positions including: Head, Pharmacy Applied Research; Head, Sterile Products, Ointment and Fluid Development; Head, Tablet and Capsule Development; Assistant Manager of Product Research and Development; and Manager, Pharmacy Research. He joined Merck Sharp & Dohme Research Laboratories in 1970 as Executive Director, Pharmaceutical Research and Development. He rejoined The Upjohn Company in 1973 and was appointed Vice President for Pharmaceutical Manufacturing.

Dr. Schroeter's research and publications have been in areas of kinetics, drug absorption, dosage form design and evaluation, steroid synthesis, oxidation, and free radical systems. He holds patents on automated analysis systems, clathrate compounds, steroid derivatives, and antibiotic dosage forms. He is the author of a technical reference book, *Sulfur Dioxide*, and a book describing pharmaceutical research titled, *Ingredient X*. Dr. Schroeter is the author of a chapter in Remington's Pharmaceutical Sciences and author of a book on management philosophy, *Organizational Elan*. Dr. Schroeter is a Fellow of the Academy of Pharmaceutical Sciences and served as its Vice President during the period 1968-69.

Dr. Anthony C. Shabica was born in Meadville, Pennsylvania. He received his B.Sc. degree at Brown University in 1938 and his M.Sc. in Chemistry from Pennsylvania State University in 1939. In 1942 he received his Ph.D. in Chemistry from Pennsylvania State University under the directorship of Professor Frank C. Whitmore and Professor Russell E. Marker. The title of his thesis was, "The Structure of the Sidechain of the Sapogenins." Upon graduation Dr. Shabica joined Merck & Company as a Senior Chemist. In 1946 he accepted a position at CIBA Pharmaceutical Company as Director of Developmental Research. He became Vice President of Development & Control in 1967, a position which he holds today for CIBA-GEIGY Pharmaceuticals. Dr. Shabica has authored and co-authored approximately 40 papers and 12 patents in the fields of steroids, vitamins, antibiotics and amino acids.

Dr. Murray Weiner was born in New York City. He received his B.Sc. degree from the College of the City of New York and his M.D. and M.Sc. degrees in biochemistry from New York University in 1943. After an internship at Sinai Hospital in Baltimore and two years in the Medical Corps, U.S. Army, Southwest Pacific, he trained in internal medicine and hematology at the Goldwater Memorial Hospital, NYU Research Service in New York. He was certified by the American Board of Internal Medicine in 1952. His professional career in the New York City area included the part-time practice of internal medicine, attending positions at several hospitals, faculty appointments at the NYU College of Medicine to Assistant Professor, and physician-in-charge, Coagulation Research Laboratory at Goldwater Memorial Hospital. He held a series of positions with Geigy Pharmaceuticals, New York, between 1952 and 1971 starting as a consultant, then Director of Clinical Pharmacology and Vice President of Biologic Research. Since 1971, he has been Vice President of Research, Merrell-National Laboratories, Division of Richardson-Merrell Inc., in Cincinnati and Professor of Clinical Medicine at the University of Cincinnati College of Medicine. Dr. Weiner has published over 170 papers and review articles dealing mostly with the clotting mechanism, anticoagulants, antiinflammatory agents, and with the physiologic disposition of drugs.

Dr. Weiner has served as Chairman of the Pharmaceutical Manufacturers Association Drug Safety Committee (1963-1966), the Gordon Conference on Medicinal Chemistry (1968), and Research Section of the Pharmaceutical Manufacturers Association (1971-1973). He has been a consultant to the AMA "New and Non-Official Drugs," the U.S. Pharmacopoiea, and has appeared as an expert witness before several congressional committees.

Dr. David I. Weisblat was born in Coshocton, Ohio, and received his A.B. Degree summa cum laude from Ohio State University in 1937. He received his doctorate degree in 1941, working under Professor M.L. Wolfrom in the field of carbohydrate chemistry. He stayed on at Ohio State doing post-doctorate research on the structure of heparin in the period from 1941 to 1943, and he joined The Upjohn Company in 1943. He was appointed Group Leader in Chemistry in 1945 and Section Head in Chemistry in 1947. For the period 1950-52 he was Manager of Chemistry and then assumed the position of Assistant Director of Research. In 1956 he was promoted to the position of Director of Chemical Research, and in 1962 he became Director of Biochemical Research. Since 1968, he has been Vice President, Pharmaceutical Research and Development, and since 1973 he has been a member of the Upjohn Board of Directors. Dr. Weisblat has authored and co-authored approximately 32 scientific papers and holds 36 patents. Dr. Weisblat's principle research interests have been in carbohydrate chemistry, vitamin synthesis, heparin structure, steroids, and prostaglandins.

Index